JOHN GRANTHAM

IRON

AS A MATERIAL FOR SHIP-BUILDING

Elibron Classics
www.elibron.com

Elibron Classics series.

ISBN 1-4021-7695-3 (paperback)
ISBN 1-4021-3330-8 (hardcover)

This Elibron Classics Replica Edition is an unabridged facsimile of the edition published in 1842 by Simpkin, Marshall, and Co., London.

IRON,

AS

A MATERIAL FOR SHIP-BUILDING;

BEING

A COMMUNICATION TO

THE POLYTECHNIC SOCIETY OF LIVERPOOL.

BY JOHN GRANTHAM, C.E.,

PRESIDENT.

LONDON:

SIMPKIN, MARSHALL, AND CO., STATIONERS' HALL COURT, AND J. WEALE, HIGH HOLBORN.

LACE AND ADDISON, STANDARD OFFICE, AND W. WEBB, LIVERPOOL; FINLAY, GLASGOW; CURRY AND CO., DUBLIN.

1842.

LIVERPOOL:

PRINTED BY LACE AND ADDISON, STANDARD OFFICE, NORTH CRESCENT.

TO THOMAS SANDS, ESQ.,

CHAIRMAN OF THE DOCK COMMITTEE, LIVERPOOL.

DEAR SIR,—I avail myself of this opportunity to thank you for having, together with our respected Chief Magistrate, honoured me by listening to so much of the annexed communication as the limits of an evening's meeting would enable me to read.

I have much pleasure in being allowed to dedicate to you this small volume, and feel that, in so doing, I address myself to one whose mind is fully alive to the importance of the subject I have here discussed, and well capable of taking a sound practical view of its merits.

The elevated position you hold, in connection with the shipping interests of this port, will render your advocacy of it of great value. May I trust that the favourable opinion you kindly expressed on hearing a portion of this communication will not be withdrawn, when, at your leisure, you have an opportunity of perusing the additional arguments with which I have endeavoured to support it.

I am, dear Sir,

Yours very respectfully,

JOHN GRANTHAM.

IRON,

AS A MATERIAL FOR SHIP-BUILDING.

AT A MEETING OF THE POLYTECHNIC SOCIETY, HELD IN THE ROYAL INSTITUTION, PRESENT, THE MAYOR, THE CHAIRMAN OF THE DOCK COMMITTEE, AND A NUMEROUS MEETING OF SCIENTIFIC MEN, MR. GRANTHAM, PRESIDENT OF THE SOCIETY, READ THE FOLLOWING COMMUNICATION ON "IRON, AS A MATERIAL FOR SHIP-BUILDING."

Liverpool, March 10, 1842.

H. DAWSON, ESQ., VICE-PRESIDENT, HAVING TAKEN THE CHAIR, MR. GRANTHAM SAID,—

MR. CHAIRMAN,

THE object, Sir, of the present communication is, to institute an inquiry into the advantages to be derived from employing iron as a material for building ships, and to compare such vessels with those built of timber.—It is not without feelings of hesitation that I venture to occupy the time of the meeting on the present occasion; not that I fear the reception I shall meet with from this assembly, past experience assuring me that every subject, involving scientific and mechanical improvement, will *here* meet with careful and considerate attention; neither is it that I fear to pledge myself to the opinions I shall this evening advance; but, that I feel I have undertaken an inquiry of deep and growing interest—an inquiry of the greatest importance to this maritime country, and, if I mistake not, to this port in particular,—and though I may, without assumption, lay claim to considerable experience in the details of the subject, it is one of such magnitude in its various bearings and ramifications, that I cannot, in entering upon it, divest myself of some feelings of anxiety lest I should fail to do it that justice which its importance demands. I should not indeed have ventured upon such a topic had my previous impressions not been strengthened, and my humble exertions stimulated, by the favourable opinion of several gentlemen on whose more mature judgment I can implicitly rely; and, even with these advantages, I must, in consideration of the difficulties I have had to encounter, and of the short intervals of leisure which my arduous occupations afford, solicit the indulgence of my fellow members in behalf of an essay which has grown upon me to an extent which I did not at first anticipate. I have sought everywhere for documents of a practical nature relating to timber and iron ships—something that I could take as the foundation of an argument—some historical view of the subject which I could carry out and enlarge. Excepting, however, works relating solely to timber vessels, or some few detached and very imperfect sketches of iron vessels—and scarcely any, on either subject, of a practical nature—I could obtain little or no assistance from published documents. I therefore determined to take an independent course, and to trust more to my own observations, and to information derived from those who would kindly assist me in collecting it, than to waste my efforts in a further and, in all probability, a fruitless search. And here permit me to thank several friends, to whom I have applied for information, for the uniform readiness with which they have replied to all my inquiries, and to state that I shall have much pleasure in receiving further additions to the facts I have already collected, from any one who will favour me by supplying them.*

Some explanation may, in the outset, be expected from me, as to the opportunities I have had of forming an opinion on this subject, that those who hear me may be better enabled to judge of the correctness of what I shall advance. A few words only are necessary on this point. My attention was first directed to the subject in consequence of my father having, in 1824, caused a small iron

---

* Since reading this paper to the Polytechnic Society, I have been favoured with several communications, which were, however, too late to be embodied into it under their proper heads. I have therefore inserted some of them as an appendix to the original paper. I beg particularly to draw attention to the letter of my friend Mr. Fairbairn, as containing the sentiments of a man of great practical and scientific information.

steamer to be built to navigate the river Shannon and the canals connected with it; from which time, a period of eighteen years, my mind has been constantly directed to the subject, and I have possessed the most favourable opportunities for examining both timber and iron vessels of all descriptions, having been engaged in building several of the latter, some of which were on a large scale. I have also been much assisted in my pursuit of this subject by my profession of Mechanical Engineer, affording me an extensive knowledge of the capabilities of iron, the mode of working it, and the proportions requisite to obtain any desired result.

After having for fifteen years possessed opportunities of thus becoming acquainted with the use of iron in ship-building, my confidence in the adequacy and superiority of that material became so fully established that, about three years ago, I determined, in connexion with other gentlemen, to engage extensively in the building of iron vessels, for which purpose we erected suitable premises at the water side, where several vessels have since been constructed.

I do not purpose, in this paper, to advance anything on the subject of naval architecture, *per se,* or to discuss the comparative merits of the forms of ships, further than as I think iron more or less suitable to carry out certain known principles, or as I conceive it may be the means of raising the character of our merchant shipping, long enthralled by an absurd system of registry laws.

In pursuing this inquiry, I shall place the various branches of the subject under their respective heads, commencing with the

EARLY HISTORY OF IRON VESSELS.

It is a common error to suppose that vessels have but recently been constructed of iron, and that the principle is only advocated by a few whose interest, as workers in iron, leads them to promote it. Many therefore, naturally enough, still view the subject with distrust, and regard it as one of the visionary schemes of this wonder-working age, which will soon be relinquished and forgotten. But I trust I shall be enabled to prove that the construction of iron vessels is not an invention of recent date; that the value of iron as a material for ship-building has long been known;

and that it has for many years been making a sure, though slow, progress towards the improved state it has already attained.

IRON CANAL BOATS.

The first traces that I can discover of the construction of iron vessels, are of those built for the canals of this country. Of these, a few, I believe, were built as far back as forty years since, and it is stated by those who have had a good opportunity of knowing, that some of them may still be in existence. During the meeting of the British Association in Glasgow, after a paper had been read on the subject of iron vessels, several gentlemen communicated facts, which had come within their own knowledge, with respect to their early introduction. A friend, in writing on this subject, states that a gentleman in Staffordshire was at that time cutting up some iron vessels which had been at work twenty-eight years. My partner, Mr. Page, was engaged in building several canal boats of iron, upwards of thirty years since; and I have myself seen iron vessels in Staffordshire, of a still greater age, but the precise date of the construction of which I could not ascertain. These facts are interesting, not only as proving that the subject has long been under the attention of practical men, but as evidence of the strength and durability of iron vessels, points to which I shall hereafter more fully allude.

THE AARON MANBY, IRON STEAMER.

The first iron steam vessel, and the first that ever put to sea, was built by the Horsley Co., for the Seine, and called the "Aaron Manby," after the name of the projector. I have lately been favoured by Mr. Manby with the particulars relating to this vessel, which are very interesting as recording the origin of iron steam vessels. He states in his communication, dated 19th February, 1842, that under a patent which he took out in France for iron steam boats, in 1820, he, with his friend Captain (now Admiral Sir Charles) Napier, formed a Society, and immediately began to construct their first boat at Horsley, but owing to some circumstances connected with the parties at Paris, she was not completed till the end of 1821. She was then sent to London in parts, and put together in dock. She took in a cargo of linseed and iron castings, and Captain

Napier took charge of her, and navigated her from London direct to Havre, and thence to Paris, without unloading any part of the cargo, she being the first and only vessel of any description that ever went direct from London to Paris. Mr. Manby continues—"Some time after, I built another iron steam vessel of the same description, with a few alterations, at Horsley; but, owing to the navigation laws in France, she could not be admitted, and was obliged to be shipped in parts, and I put her together at Charenton, near Paris, where I had then established iron works, and where I subsequently constructed two other iron steam boats, the whole for the navigation of the Seine. They continued prosperously at work till 1830, when, owing to the revolution, and some disputes among the shareholders, they were sold to a new society. In this new society I had no further interest, but they continued navigating up to the period of my quitting France, and I believe are all at work at the present time. From 1822 to 1830 the hull of the Aaron Manby never required any repairs, although she had been repeatedly aground, with her cargo on board."

This vessel does not appear to have been very fast, but excited considerable attention, as will be seen by the following extract from *Le Constitutionnel*, of 13th of June, 1822:—"Le bateau à vapeur en fer, l'Aaron Manby, Capt. Anglais Napier, est arrivé hier, Lundi, à huit heures du soir au Port St. Nicholas, avec un chargement de graine de tréfle, qu'il avait pris à Rouen, et de quelques pièces de fonte et de mecániques, venant d'Angleterre. Le bateau Français à vapeur, Le Duc de Bordeaux, arrivé de Rouen samedi au soir, avec un chargement complet qu'il n'avait pu mettre à terre en entier, était parti du port à quatre heures du soir pour aller à la rencontre du bateau Anglais qu'il a atteint à la hauteur de Saint Cloud, en face des cascades. Ils sont partis ensemble, de la pointe de L'Isle Seguin, et le bateau Français, dont la manœuvre est visiblement supérieure, est arrivé au Port St. Nicholas quarante minutes avant l'Anglais. Les curieux ont été, pendant toute la journée, visiter les deux bateaux.

IRON STEAMERS ON THE SHANNON, &c.

The next iron steam vessel with which I am acquainted was one built by the Horsley Co., under my father's superintendence.

This vessel was commenced about the year 1824; was put together in this port, and, after a series of delays, crossed the Channel in 1825, and proceeded to her destination, Lough Derg, on the River Shannon. Since then she has been constantly at work, and is now in good condition. This vessel was the origin of the extensive and spirited Company which now occupies that splendid river, and which is conferring important benefits on the large tract of country through which the Shannon flows. My friend Mr. Williams, Managing Director of this and of the City of Dublin Steam Packet Company, at this time directed his attention to the subject of iron vessels. With his usual discernment he foresaw the advantages to be derived from this source to steam navigation, and has ever since been a steady advocate of the principle. It was at his recommendation that the Shannon Company continued to construct iron steam vessels, and there are now six at work, all in excellent condition, two of them in salt water. Iron vessels now began to attract the attention of those who were engaged in river and canal navigation, and several were built in this country and on the continent. The first that were built in this town, were undertaken by Messrs. Fawcett and Co., and under the superintendence of Mr. Page.

IRON VESSELS BECOME GENERAL.

Shortly after this time, Mr. John Laird, of North Birkenhead, commenced building them on a large scale, and has since been extensively and successfully engaged in this pursuit. Mr. Fairbairn, of Manchester, also very early took an interest in iron vessels, and was a party to a series of experiments made at Glasgow, in which iron vessels were employed.

During this period, and for some years previously, I watched with the greatest interest the progress of iron vessels, feeling convinced that they must ultimately supersede those built of timber. Many others have now commenced the business, and numerous iron steamers and iron sailing vessels of large tonnage are now afloat, or building. Great numbers of iron steamers are plying on the Thames,† the Mersey, the Clyde, and on nearly all the continental

---

† The principal builders in the Thames are Messrs. W. Fairbairn and Co., and Messrs. Ditchburne and Mare.

rivers. Some of a larger and stronger description are regularly employed in making sea voyages with heavy cargoes; these latter fully answer the purpose for which they were intended, and are highly satisfactory.

The Iron Steamers running between this port and Glasgow, and built by Messrs. Tod and M'Gregor, are well known. Of these the Princess Royal is the largest, being upwards of 800 tons. She is a fine vessel, of immense power, and is unequalled in speed as a sea-going steamer. The builders have favoured me with some interesting facts connected with iron vessels, of some of which I shall hereafter avail myself. Iron steam vessels are even now fighting our battles in the East, among which the Nemesis and Phlegethon, built by Mr. Laird, are entitled to particular notice, from the prominent part these vessels have taken in the Chinese war.

Of iron sailing vessels, some have made voyages both to the East and West Indies; and I entertain the hope that, by means of a clear statement of facts, which my attention to the subject has placed in my hands, and a description of the mode of construction, and of the advantages possessed by iron, as a material for ship building, I may be instrumental in correcting the false impressions now abroad on the subject, and promote an object, which I believe to be of the utmost importance to our national and commercial interests. The Ironside was the first iron sailing vessel of any magnitude that was employed for sea voyages, and she has been highly successful. She has made three trips to and from South America, and fully realizes all the advantages proposed in her construction.

A proposition was made in 1838 by Mr. Holmes, C. E., to construct large iron steamers for the communication, *via* the Red Sea, to India. Some of his calculations as to the profit to be derived from this speculation are very doubtful; but his general views appear to be correct. He certainly deserves credit for his exertions, which, however, seem to have been unsuccessful. His report would receive much more attention now than it did at the time of its publication, as it is very evident that he judged correctly as to the comparative merits of wood and iron vessels for long voyages; and a great change of opinion is taking place on the subject. Several men of experience gave evidence in favour of iron vessels before the committee appointed to investigate the subject. I shall have occasion to quote from Mr. Holmes's report in my subsequent observations, as it accords very much with my own views.

Among the best publications relating to iron vessels was a paper written for the *United Service Journal*, in May, 1840, by Mr. Creuze, of her Majesty's Dock Yard, Portsmouth, and author of the able Article on Ship Building in the last edition of the *Encyclopædia Britannica*. This communication was suggested in consequence of the injury sustained by the Nemesis, on the rocks off Scilly. Having put into Portsmouth for repairs, she was there seen and examined by Mr. Creuze. I shall occasionally submit extracts from this paper, it being the production of a gentleman who evidently possesses a scientific knowledge of his subject, and is fully alive to all the difficulties which are to be encountered in naval architecture.

The iron steamer now building at Bristol is one of the boldest efforts of modern times, and forms a remarkable feature in the history of iron vessels. I shall give a short description of her towards the close of this communication, as furnished to me by Mr. Guppy, the talented managing director of the works.

CONSTRUCTION OF IRON VESSELS.

With a view to the better elucidation of the subject, I now proceed with a description of the ordinary method of constructing iron vessels. In doing this, it may be considered, by those engaged in the same pursuit, that I am laying open to the public eye the secrets of our business, and encouraging others to enter into competition with us. But the principle of iron vessels is comparatively new; there is still much room for the exercise of ingenuity and practical skill; and I consider it to be necessary that the public should be made acquainted with the general principles on which we proceed, as the best means of giving confidence to the shipowner and extending the demand for vessels of that material. I am convinced, moreover, that it would conduce to our mutual interest to communicate to each other more freely the results of our experience, in order to insure as much as possible the introduction of every improvement, and thereby promote the employment of iron-built ships.

I shall describe only the *ordinary* mode of construction, as now generally adopted, and not attempt to suggest any new plans; although it is easy to conceive that the subject admits of much improvement, as the class of vessels to be built of iron increases in size.

KEELS.

No uniform plan has yet been adopted for the keels; and they are made of various forms, according to the object to be attained in the respective size or adaptation of vessels, or the opinions of the builders.

Figure 1, Plate 1, is the form generally adopted for steamers, and is favourable where the depth should not exceed three or four inches.

Figure 2 is a plan frequently applied to vessels, but I consider it objectionable; the weight of the ship when taking the ground resting on the rivets at *a*, *a*.

Figure 3 is a plan adapted to sailing vessels where a considerable depth is required. All these keels form channels into which the bilge-water drains, and from which it is pumped.

Fig. 4 represents a plan which I conceive to be an improvement upon No. 3, the bottom plates being carried across and bent down at *a*, so as to form a limber, and the keel being placed outside, so that, should it chance to be broken by striking on a rock, or otherwise, the water cannot enter into the hold. The plan of bolting on a false keel of timber to an iron vessel cannot be too highly deprecated. It nearly caused the loss of "The Iron Duke," a vessel of 350 tons, which, striking heavily on a sand-bank, tore away the false keel, and, breaking off the bolts by which it was fastened, the water entered by the holes. Luckily, they were easily stopped; but, probably, no damage would have accrued if the whole had been of iron.

STEM AND STERN POSTS.

Various plans have been adopted in constructing the stem and stern posts of iron vessels. In many cases the stem has been formed by a plate so bent as to have the outward appearance of the stem of a wooden vessel *(see Fig. 5 ;)* but I consider this plan to be defective and dangerous. The stem of a vessel so constructed is liable, should she strike any hard body, to become indented; and, from the injury necessarily sustained by the plate in forging, it is liable to crack. A better and much simpler plan is now generally adopted, *(see Fig. 6.)* A bar—a piece of strong flat iron—is fashioned to the required form of the stem, extending a few feet along the keel. The plates which form the sides of the vessel extend to the fore part of this stem, and the whole are firmly rivetted together. A new vessel lately built by us with the stem so constructed, ran straight on to a stone pier at considerable speed; and although a large stone was displaced by the shock, scarcely any traces of the accident could be found on the vessel.

The stern post is not so easily managed. The necessity for making it as broad as the rudder stock requires a very thick bar of iron, if it is to be made solid, and it is not therefore often attempted so to construct it. I, however, prefer the same plan as that recommended for the stern, as I consider it best suited to resist the severe blows to which vessels are often exposed when crossing bars in a heavy sea, or running stern on to solid bodies, should they part their cables.

FLOORINGS.

The floorings are first formed by a single or double bar of angle iron, *(see Figure 7,)* to the vertical side of which a plain strip of iron, suiting the shape of the bottom, and extending from one bilge to the other, is rivetted; and on the upper edge of this strip a piece of angle iron is fastened, one side placed horizontally, to form a bed for the ceiling and kelsons. It is of great importance that the frames should not be too far apart, (say 15 to 18 inches,) and of considerable strength, the additional weight of the bottom being no objection, and great firmness being necessary to preserve the vessel should she take the ground.

SIDE FRAMES.

The side frames, or ribs, are formed of angle iron, extending from the deck, and over-lapping the angle iron of the floorings at the bilge, a length of 2 or 3 feet. To each, or to every other one of these ribs, another piece of angle iron is rivetted in reverse position, as shewn by Figure 2, Plate 2. To this last piece of angle iron the wood ceiling is bolted.

GUNNELS.

On the top of the ribs is placed a strong bar of angle iron running round the vessel, and taking the shear of the deck. This plan has not been generally adopted, excepting in Liverpool. The object is to assist in forming the gunnel, and I consider it important in giving additional strength, and in securing the wood-work to the iron. To this angle iron has lately been added a plate 12 or 15 inches wide, placed horizontally and extending under the deck, as shewn at *l*, Plate 5. This plan affords several advantages. It gives additional security in making a water-tight and strong joint at the edge of the gunnel, where the wood-work joins the iron.

MODE OF PLATING.

The plating is the next subject for consideration; and in this, as in other portions of my description, I do not purpose to point out the proportionate scantling required for different vessels, an attempt that would involve lengthened calculations; but rather to confine myself to general terms, and to state dimensions so far only as is necessary to convey some idea of the usual methods adopted in this principle of building, for the information of those who may not have an opportunity of investigating the subject for themselves.

Plates of all forms and dimensions have, at various times, been used by different builders, from 1-16th to 3-4ths of an inch in thickness, according to the size and required strength of the vessel. These are not, however, the limits within which the plates must of necessity be made: some of much greater strength will probably, in the further progress of the art, soon be in requisition. It is, I believe, generally acknowledged that until of late the plates used for iron vessels have been much too light; but, as the subject becomes better understood, plates of greater strength are employed. On this branch of the subject, a much greater difference of opinion is manifested than on that of "framing;" the object of the latter being principally to preserve the form of, and to give stiffness to, the former. The impolicy of using light plates in the construction of a vessel is sufficiently obvious, for a very few additional tons make an almost imperceptible difference in the draft of water, but materially increase the strength and durability of the ship. The scope which is afforded in iron vessels for increasing the strength of the framing is almost unlimited, and this by simple and evident means; but the question as to the best disposition and proportionate strength of the plates, involves so many difficult and nice enquiries, that experience alone can satisfactorily determine it.

JOINTING.

There are two methods by which the plates are united. One of them *(see Fig.* 1 *and* 2, *Plate* 3*)* is, by bringing the edges of the plates together; a narrow plate, or strip, being placed inside to connect them by rivetting, thus forming a flush joint externally. The other method *(represented by Fig.* 5 *and* 6*)* is, by lapping the edges of the plates on each other, and rivetting them together. The former method corresponds with the carvel build, and the latter with the clencher build, of timber vessels. In both cases the position of the plates is longitudinal with the ship, as in wooden planking.

Plates united with flush joints will evidently sustain the greatest strain arising from a force applied to or affecting their horizontal edges, for the weight is entirely removed from the rivets, and the supposed tendency to tear open those parts of the plates which are weakened by the rivet holes, will only apply to the vertical joints. With the lapped joints, on the contrary, nearly all the strain produced by a weight applied to the edge, or rather margin, of the plates, is thrown upon the rivets with an apparent tendency to cut or shear them off in the middle of their length, or, in other words, in the centre of the lap; excepting in so far as this strain is counteracted by the frames, (corresponding with the frames of a wooden vessel,) which, at short distances apart, cross those joints.

In order to decide which of these two methods is the most advantageous, it becomes necessary to consider the nature of the strains to which vessels are liable. In taking the ground, or in a heavy sea—the cases in which vessels are most tried—the strain on the bottom arises from an upward pressure; while, on the sides, two forces operate, one given out in a vertical direction, similar to weights on a beam, and the other (operating only when the vessel is afloat) from the inward pressure of the water. From these two forces arise all those compound actions which cause the straining and wrenching to which vessels are exposed.

When a ship labours in a heavy sea, the strains are momentarily varying in amount and direction; it is therefore impossible to establish any rule by which their effect may be counteracted, and the only safe course is to apply the maximum strength that experience has shown to be requisite, or of which the materials employed in building, are susceptible. The power of an iron plate, while it is kept perfectly straight, to resist a strain or pressure applied to its edge, is very great; and as this power increases in proportion to the square of the depth, we can easily estimate the strength of a ship's side to resist vertical strains, when it is formed, as it were, into one large plate. I believe this, indeed, to be so far beyond what can ever be required, that we can readily forego a portion of it, to obtain strength in another direction,—that is, to resist strains which tend to disturb the correct position of the plates. And although the strains arising from the vertical pressure on a ship's side very much exceed those caused by the inward pressure of the water, or by the ordinary concussions of hard bodies, it can easily be shown by calculation that the strength of the plates in a vertical direction is proportionately much greater to resist strains given out in that direction, than their strength in a lateral direction to resist strains arising from inward pressure. From this reasoning, the lapped joint appears to be preferable to the flush joint; and it is probable that if the former principle had invariably been adhered to, much trouble and expense would have been saved, as the flush plating seems to have originated in a mistaken idea that it increased the strength and improved the appearance of the ship.

SINGLE AND DOUBLE RIVETTING.

As before stated, the vertical pressure exerted on the lapped joint has an apparent tendency to cut off the rivets in the centre of their length; and a very general impression prevails that, if a great strain were applied to the lapped edges of a number of plates, such would be the result. Experience, however, does not bear out this supposition, for it is well known to practical men that the plates themselves almost invariably give way first, tearing in the parts weakened by the holes formed to receive the rivets. As, however, the space between each hole, in single rivetting, exceeds the diameter of the hole, more than *half* the original strength of the margins of the plates overlapped remains after the holes are punched; and where a double row of rivets is adopted, more than *two-thirds* of the original strength of each plate is left.

To render this clearly understood, I should observe, in the first place, that the system of double rivetting adds very much to the strength of the joints, and for this reason:—It is an established rule to use a certain number of rivets to each lineal foot of the joint, and these are considered necessary to draw the plates firmly together. In the single row, the rivet holes so nearly approach each other that the strength of the plates is much reduced; but, by the double row, not only can the same or a greater number of rivets be used per lineal foot, but a greater space is left between the holes and an increased stiffness is imparted to the plates at the joint. This will be seen by reference to Figures 5 and 7, Plate 3, the line *a b*, in each figure, showing the part most liable to break.

In an interesting paper on the strength of iron plates, as applied to ship-building, read by Mr. Fairbairn to the British Association, at Glasgow, that gentleman states that, assuming the strength of plates to resist tension to be 100
A double rivetted joint will be......................... 70
And a single ditto will be......................... 56

In the system of flush jointing, twice the number of rivets are required; and, therefore, to obtain the advantages of double rivetting, four rows of rivets are necessary to each joint. *(See Fig. 3 and 4, Plate 3.)*

FURTHER ARGUMENTS IN FAVOUR OF LAPPED JOINTS.

In further considering the subject with reference to the strains arising from the lateral pressure of the water, or from internal concussions, the comparison will be much in favour of lapped joints. When plates rivetted with flush joints are driven inwards, their edges *(see a, Fig. 9, Plate 3)* become fulcra to the powerful leverage of the plates. A great strain is thus thrown on the rivets, tending to tear or disrupt the plates in the direction of the rivet holes, as before noticed. In the lapped joint, no such leverage can arise; but the whole will yield together, and, although the plates may bend, the joints will remain perfect. *(See Fig. 10, Plate 3.)* A remarkable instance of this fact is furnished by the ship "Vulcan," which struck on a rock, by which the plates were deeply indented, without causing any leakage from the joints. I shall have occasion hereafter to allude more fully to this circumstance. The lapped joint also gives more effect to a property which bodies possess of uniting, when firmly pressed together, the more prominent parts of the one sinking into the indentations of the other. I conceive this to be a great assistance in preserving the strength in the horizontal joints of a ship's side.

There are other practical difficulties that arise from flush joints, amongst which may be named that of caulking. This operation is performed by striking the edges of the plates with a sharp tool, and forcing the metal into the joint, entirely closing every part which the rivets had not previously rendered tight. This fills up every crevice, without the aid of cement, and is evidently more effectual in the lapped than in the flush joint.

SOME OBJECTIONS TO LAPPED PLATES ANSWERED.

Objections have been made to the lapped joints, on the ground that they give a rough appearance to the hull, and impede the sailing. These views of the question appear to me to be erroneous. There is a great degree of lightness and elegance in a clencher-built vessel that cannot be imparted to a smooth-sided carvel-built ship. The edges of the planks, or the plates, in the former, present to the eye a pleasing series of lines, in gentle curvatures, showing the shear; and, with respect to sailing qualities, it is very questionable (provided the shear of the planks, respectively, be not too great) whether the edges, presented in almost straight and parallel lines, at all impede the vessel's velocity, while they render her stiffer and less leewardly by giving her, like so many keels, a better hold of the water, and, at the same time, make her more lively in a sea-way. Many of our fastest pleasure-yachts are of this build, as were most of the continental smuggling luggers, during the French war, which, excepting in heavy gales, frequently escaped our fleetest frigates. If, then, the objection to the clencher-build be trivial or untenable, as regards wooden vessels, how much less will it hold good with respect to iron vessels, in which the edges of the lapped plates are not proportionately of half the thickness of planks.

If the objections hitherto entertained to lapped joints be satisfactorily removed, and the advantages of the system be acknowledged, another argument in their favour may be admitted, namely, that they save both weight and expense.

RIVETTING.

The rivets, as before stated, seldom give way; but the plates, should fractures take place, would crack in the line of the rivet-holes. It is, therefore, of great importance that the holes should not be unnecessarily large, or too close together, and that the strength of the rivets and the plates they fasten should have a due relation to each other. To avoid exposing the projecting head of the rivet, the outer hole is counter-sunk, and the rivet is of such a length that, when hammered down, it exactly fills the hole, and leaves a smooth external surface. The counter-sunk head is also a great safeguard against the effects of corrosion; for the plate must become nearly worn through, before the rivet loses its hold. This will be seen by reference to the drawings.

Figure 11, Plate 3, represents two plates rivetted in the ordinary way, as in boiler-making, in which the projecting heads on both sides of the plates is no disadvantage. The form of the rivets, before being hammered, is shown by the dotted line.

Figure 12 represents the plates with the outer hole at *a*, counter-sunk, the portion of the rivet filling it forming a substitute for the usual head.

Figure 13 exhibits the usual form of the holes when counter-sunk; but it is found that the rivet sometimes breaks at the dotted line *a, b,* by being hammered after it is cold.

The enquiry is frequently made, whether there is not great danger, should the heads fly off and the rivets fall out, of lessening the strength of the vessel, and leaving a hole for the admission of the water. The answer to this question is that I never heard of such an accident. For a few years, all the repairs of the iron vessels in the Mersey were done at the establishment with which I am connected; yet, during that time, no occasion was found to replace a single rivet. The heads occasionally break off while the workmen are in the act of rivetting, and when the holes are badly formed; but when made as shown in Figure 12, this seldom occurs. Should, however, the head of a rivet break off when the vessel is afloat, it by no means follows that the rivet itself would fall out. On the contrary, when originally secured in a workman-like manner, the piece remains so firm in the hole that it requires the application of a steel drift and a heavy sledge to drive it out. This circumstance is easily accounted for by an examination of the transverse section of the rivets, *(Plate 4)* as they fill the holes in the two plates; and the holes being seldom exactly fair with each other, the rivets swell out and accommodate themselves to the precise shape of the holes, like melted lead run into a mould. The annexed drawing, taken from some rivets, which were cut through the centre in the planing machine, explains this peculiarity.

DECK BEAMS.

The deck beams, in iron vessels, are now generally made of iron, and afford another proof of the adaptation of that material for ship-building. The plan generally adopted is to place two pieces of angle iron back to back, with a strip of iron between them. To the horizontal planes of these angle-bars, the plank is bolted, either by wood screws or bolts and nuts, as shown in Figure 7, Plate 2. The wood screws are sent up from below, and have the advantage of not injuring the surface of the deck, to which they need not necessarily rise. Bolts and nuts are, however, more secure.

The angle-iron of the deck beams may be continued down the vessel's sides two or three feet, and be rivetted to the plates; and in the angle thus formed, an angular plate is introduced, as shown in the sketch, *(Fig. 6,)* forming a most efficient knee, of comparatively little weight.

The lower deck beams are similar to those of the upper deck; but an iron stringer is generally added, consisting of a strong plate, or a stout angle bar, rivetted to the side-frames, and to the beams, uniting the whole firmly together. *(See parts marked l, m, Figures 1 and 2, Plate 2.)*

BULKHEADS.

Iron vessels admit of the very easy introduction of iron water-tight bulkheads. These are formed by plates, rivetted to the ribs or frames, both on the sides and bottom, each bulkhead making a complete transverse section of the vessel, and the whole being so secured as to prevent water passing from one side to the other. Several of these being introduced, divide the vessel into water-tight compartments.

The bulkheads affording the greatest protection are those placed a few feet respectively from the stem and stern; the forward one checking the water that would enter through a damaged stem, and the after one averting the danger of any accident that might arise to the stern-post, or rudder-braces. The water received into these small compartments would very slightly impede the way of the ship by throwing her out of trim, as the quantity they would contain would be comparatively trifling. The bulkheads more a-midships are principally of service in giving strength to the vessel, but also afford safety in the event of fire, and prevent it spreading beyond the compartment in which it commenced. An ordinary leak, too, in one of them, may be overcome, and no damage arise to the cargo or stores in those adjoining. These are advantages which it would be difficult to attain so effectually in a timber-built vessel.

THE USE OF IRON EXTENDED TO OTHER PARTS.

The enumeration I have made embraces the principal parts of vessels in which iron has been substituted for wood. The remaining parts of what are called iron vessels, from the shell being of that material, such as covering-boards, waterways, decks, stancheons, bulwarks, cabins, masts, &c., are, as yet, of wood, with little alteration from timber-built ships, excepting in so far as is necessary to secure them to the iron. I am far, however, from thinking that iron is applicable to those parts only of a vessel to which its use is now confined. On the contrary, I am of opinion that decks, bulwarks, and other parts will ere long be made of that strong and durable material—a work by no means difficult of accomplishment, and which I feel convinced would much increase the security of the vessel and cargo. Even iron masts have, in more than one instance, been adopted; and one has been in use for upwards of twelve years in the "Ballinasloe" steamer, without sustaining any perceptible loss in its strength. This mast also serves as ventilator to the cabin. My present object, however, is not to enter into details of untried plans, but to confine my remarks to the ordinary system of building with iron, as at present pursued. To those who have had opportunities of seeing and inspecting iron vessels, the particulars furnished will be familiar; but to such as have not had this advantage, the brief description given will be found useful in enabling them to follow me through the next division of my subject.

IRON VESSELS CONSIDERED AS A COMMERCIAL QUESTION.

Under this general head, I purpose to submit some remarks on the peculiarities of iron vessels, and institute a comparison between them and timber-built vessels, embracing those features in each that principally affect our commercial interests.

What, then, are the points which it is the province of the builder to study and attain? What are the objects most desired by the merchant in the choice of a ship? These, I conceive, are—

1 Strength, combined with lightness.
2 Great capacity for stowage.
3 Safety.
4 Speed.
5 Durability.
6 Economy in Repairs.
7 Cost.
8 Draft of water.

I trust I shall be enabled to prove that iron vessels possess advantages under all these heads in so eminent a degree as to render them superior to wooden vessels; and shall address myself to each point in its respective order:—

FIRST—STRENGTH COMBINED WITH LIGHTNESS.

This subject involves two considerations, the strength of the materials and the mode of uniting them.

The great strength of malleable iron to resist strains in every direction is well known; but, to those who are not conversant with the subject, the extent to which this advantage may be carried is not at first apparent; or how the material may, from comparatively smallpieces, be so combined in large masses, as to form the ponderous body of a ship; and they are thus too apt to prescribe a limit to its use. An opinion, indeed, is now very generally entertained that iron may be suitable for small craft, but is inadequate for the construction of vessels of heavy burthen. This, however, is a supposition so erroneous that the reverse would be much more correct; for large vessels will afford the best practical demonstration of the superiority of iron for ship-building. In the application of timber, obstructions increase in a ratio proportioned to the increased size of the vessel to be built. How often has the ship-builder the greatest difficulty in obtaining timber to suit the varied curves of our finest ships! How often is the country despoiled of its noblest ornaments by the tempting prices he is compelled to offer for its most magnificent oaks, the largest of which are frequently insufficient for his purpose! How are his brains racked, and his patience tried, in seeking for crooked timber necessary to frame a sharp floor, or a square bilge! How often is he obliged, though he knows it to be injurious, to scarf the frames, for which no timber can be found sufficiently large to enable him to avoid such defects! And is not this one cause, amongst others, why our building yards are empty, while our ports are filled with ships from other nations, in which timber is more plentiful and the choice more extensive!

But how stands the case when we turn to iron? Where is the frame, even of the most intricate form, that our smiths cannot mould? Where the frame or beam so large that iron cannot be found of which to fashion it, and that, too, if need be, without a scarf? Here there are no knots, no sap, no cutting across the grain. Here there is no useless timber, placed merely to *fill in,* or to cross buts. Here every inch of material is of service, and every scrap applied to some useful end.

Iron has great power to resist tension in any direction, as is shown by Mr. Fairbairn, in the communication before alluded to, in which he proves that scarcely any difference in its resisting powers is to be perceived, whether the strain be applied in the direction in which the plate is rolled, or across it. Iron has also, to a high degree, the power of resisting compression. Timber, it is admitted, has great power to resist tension in the direction of the grain, but is very deficient in strength across the grain; and its power to resist compression is also very limited, especially when it is exposed to moisture. Again,—timber, after being some time in use, becomes brittle, and is but little disposed to bend. Good malleable iron, on the contrary, may be bent double even when cold, and does not become brittle with age, except when converted into an oxide. The ease also with which iron beams and frames can be wrought, and the facility of obtaining them of any dimensions in one piece, overcomes one of the greatest difficulties in ship-building.

I have before stated, that the power to increase the stiffness of the hull, when built of iron, is unlimited: and provided the shell has originally been made sufficiently thick, additional strength may, at any time, be given to the frame. Of this fact we have a remarkable instance at present under our observation, in the large iron ship "John Garrow," now undergoing alterations at our works, and of which I shall hereafter furnish some details.

The strength of beams to resist vertical pressure increases as the square of their depth. The advantages to be derived from this may be simply illustrated. By making a beam 17 inches deep instead of 12, we double its strength, but its weight is increased in the proportion of only 17 to 12; or by doubling the depth of the beam, we increase its strength four-fold, but we only double its weight,—and so on to any extent.

As we have sufficient proof that the utmost strength that can be required in vessels may be obtained from iron plates, it only remains to dispose of them in the position in which their strength is found to be most available. If the shell be sufficiently strong to resist all the strains arising from tension and compression, it is only necessary to stiffen it so much by adding other plates perpendicular to those of the shell that it shall not deviate from its correct or original form.

If we view the subject with respect to very large vessels—ships of the line or first-class steamers, in which intermediate decks are not only not objectionable, but requisite for the guns and stores, and in which entire or partial bulkheads may be multiplied without causing much inconvenience—we have in iron the means of dividing the shell into small sections, affording the strongest supports directly opposed to the strains that tend to disturb the form of the plates, and in the direction in which they can least resist such strains. The sides between these decks and bulkheads would be strengthened by the ordinary mode of framing. In addition to this, the lower part of the vessel might be again intersected by longitudinal divisions; and these decks, bulkheads, and subdivisions would not only be securely fastened to the shell, but to each other,—the whole thus becoming a mass of almost irresistible strength, meeting the strains in whatever direction they might arise, while, at the same time the weight would be far below that of a timber vessel of only moderate scantling, and the room for stowage would be much greater.

FASTENING.

A comparison of the different modes of fastening timber and iron vessels, and uniting their separate parts, is as favourable to the latter as the question we have just considered. We will take, in the first instance, the case of an expensively copper-fastened timber-built ship. Copper and iron bolts are used to secure the frames to each other, extending in some cases through the kelson, floorings, and main keel, and in various other parts. But what generally are these bolts? They are plain bars or rods of copper, driven into the holes prepared for them: a washer is then slipped over the other end, and the carpenter with a small hammer forms on each a kind of head, pressing the washer firmly down into the timber. All this is very well so long as no very severe strain takes place, but is found to be wholly insufficient for the binding together and support of large vessels. Of this fact we have abundant proofs. I have myself seen a vessel, nearly new, put into the graving dock with several of her bolts projecting out two or three inches from her bottom, they being started or drawn, and rendered quite useless by the working of the ship.

A most elaborate system of bracing by iron straps, knees, and bolts with nuts, in addition to the ordinary knees, diagonal internal planking, stringers, &c. is now pursued. Tree-nails used to secure the planking are giving way to bolts, either of copper, iron, or composition, and great additional strength is thus imparted. But the grand objection still remains. The bolts do not accord with the materials they fasten together, and the wood is so soft that when a severe strain arises, a general yielding takes place, by the bolt heads sinking into it, and the wood itself giving way to the pressure it receives from the neighbouring parts. In proof of this it is only necessary to examine the copper sheathing of a large vessel after a trying voyage, when the waving surface too often betrays the defects within. This effect of straining is most perceptible in steam vessels, particularly under the paddle beams, and sometimes abaft them. The creaking of a timber-built ship, too, so distressing to landsmen, and adding so much to the terrors of a storm at sea, is the result of the unceasing motion which takes place in the seams, acting on the internal timber work, and particularly on the cabin partitions and bulkheads.

A wooden vessel is made up of a great number of parts of various forms and sizes, placed in different directions, which, like the stones in an arch, are dependent on each other for support. If one becomes loose, the whole structure is more or less endangered. This is evident from the fact that much of the stiffness of a wooden vessel depends on the caulking of the outer plank.

I may add, in corroboration of these facts, that it is the general custom with builders to leave the gangways of the bulwarks in wooden ships unfinished, lest the hull should so much alter its form by settling, in launching, that the rails would not again fit their places ; and no builder would willingly copper a vessel when new, but rather allow her first to find her own position in the water, as she would then be less liable to wrinkle the sheets. The defects noticed are indications of weakness and tendency to straining, but they are so common, and, at the same time, inseparable from the construction of ships, that no one deems them worthy of notice as objections; and I perhaps expose myself to the chance of being considered an alarmist by merely alluding to them. My only object, however, is to state the truth, and to leave others to judge for themselves.

Mr. Creuze, (to whom I have before alluded,) in his paper on iron vessels, thus reasons:—" Two questions naturally arise; first, What are the advantages or disadvantages of the substitution of iron for timber, in the construction of ships?—and, secondly, to what limit may this substitution be advantageously carried? Amongst the advantages are the employment of a less costly material, of which the supply is inexhaustible, and for which supply we are totally independent of other nations; also, the greater durability of the material, not only from its relative durability with that of timber, but from its requiring no such metallic sheathing to protect it from the ravages of worms. Also the greater durability of the structure, as a whole, in consequence of the greater permanency in the perfect combination of its several parts, arising from the fastenings being of the same texture as the portions of materials brought into connexion. The metallic fastenings to a timber-built vessel act, it must be remembered, not only chemically, but also mechanically, to accelerate her destruction so soon as the close connexion of the several parts is at all diminished."

The objections arising from the use of fastenings of a material so totally different from that of which the hull is composed, are entirely removed in iron vessels. In the first place, the outer shell of the vessel is composed of a series of plates, so rivetted together that its strength is nearly equal to what it would be were it possible to form the whole of one plate. This shell is independent of all indirect means for preserving its completeness. It forms one grand whole, of the same material throughout, and that of the strongest kind. This shell is stiffened, as before described, by ribs crossing the joints of the plates at short distances apart, and giving an additional security, Beams, knees, bulkheads, all are brought together in one firm mass, and united by numberless short unyielding rivets. I may venture, indeed, to say, that more real serviceable fastening is often employed in the space of a few inches, in an iron vessel, than is in most instances brought to bear on one entire beam of a timber-built ship.

In proof of the superior strength of iron vessels, I need but adduce a few examples that have fallen within my own observation. A friend (before alluded to) in speaking of the iron vessels used on canals, states that an extraordinary difference is found between the relative strength of wooden and iron vessels; that the latter are employed in the severest service, such as carrying iron stone, pig iron, and refined iron from the blast furnaces, together with other heavy and unwieldly articles; and he concludes by expressing his conviction that "they will stand anything for which canal boats can be used."

I myself witnessed an accident that occurred about fifteen years ago to my father's small steamer, and which would probably have caused the immediate destruction of a wooden vessel so circumstanced. While on her accustomed voyage through Lough Derg, sailing with a fair wind, and running probably seven knots an hour, she came in contact with the sharp point of a rock, up which she ran about one third of her length, raising her head considerably out of the water. We blew the water out of the boilers, and lightened the vessel as much as possible. A large number of the country people put off to our assistance, and, an anchor being dropped, assisted at the windlass to heave her off, but we were unable for a long time to move her. When we again got her fairly afloat it was found that she was perfectly tight, and no further notice was taken of the circumstance until about twelve months afterwards, when she was put into the graving dock. I then endeavoured to find the place where she had struck, but could only discover a very slight and unimportant dinge. The "Lady Lansdowne" iron steamer, lately ran on the rocks in the same lake, but received no perceptible injury. The case of the "Garryowen" iron steamer, in the estuary of the Lower Shannon, during the great storm of the 6th of January 1839, has before been published; but may here be repeated. She broke from her anchors; went ashore near Kilrush, and beat with great violence for some hours, on a hard beach. She, however, sustained no damage, although, out of twenty-seven vessels, in and about the same neighbourhood, not one other escaped serious injury, and some were entirely lost! In a Report, laid before the House of Commons, on the accidents to steam vessels, the following account is given of the stranding of the "Garryowen":—"We went ashore about two cable-lengths to the eastward of the pier, (Kilrush,) and struck very heavily for the first hour. The ground under our weather bilge was rather soft clay, covered with shingle and loose stones, some of them pretty large. Under our inside or lee bilge, the ground was very hard, being a foot-path at low water. I was greatly afraid she would be much injured in her bottom, but I am happy to say she has not received any injury: in fact, her bottom is as perfect and as good as on the day she left Liverpool,—*not a single rivet started, nor a rivet head flown off.* If an oak vessel, with the cargo she had on board, were to go on shore where the "Garryowen" did, and get such a hammering, they would have a different story to tell. Out of twenty-seven vessels that got ashore that night, the "Garryowen" is the only one that is not damaged more or less."

An iron sailing ship, called "The Vulcan," met with a trying accident, the particulars of which I shall give in the Captain's own words:—"As we were taking the sound, she struck on a reef of rocks, with only three feet water, at the same time the ship was drawing six and a half feet, and going at the rate of five-and-a-half knots; but, owing to the wind being aft, she forced her way over the reef without receiving any injury, as she made no water afterwards. My opinion is, that, if she had been a wooden vessel, her bottom would have been greatly injured, and very likely she would have sunk. During all the different trips the "Vulcan" made in the course of the ten months I belonged to her, I can with confidence give her the character of being a very fast-sailing, powerful, and excellent sea boat; far surpassing any wooden vessel I had ever been in before. We have at different times been in company with other vessels lying-to and fore-reaching under close reefed sails, and the "Vulcan" carrying double reefed top-sails and jib half-boom out. She gives great satisfaction to her owners, both as regards sailing fast, carrying a great cargo, and going on a light draft of water." When this vessel was in the graving dock last year, I saw the result of the blow the "Vulcan" had sustained. The plates were indented in the fore body, where she first struck, and slight marks were made in several parts along her bottom, caused by her continuing to strike as she passed over the reef. No plates were cracked, no rivets broken, excepting a few which had held one of the ribs. These were readily replaced, and all was again perfectly sound.

The "Iron Duke," a ship measuring about 350 tons, affords a favourable specimen of the strength of iron vessels. She had been one voyage to Calcutta. On her return thence it was reported that she steered badly, and her compasses were not correct. To obviate the first evil, we were employed to make some alterations in her, which, the captain afterwards reported, considerably improved her; but, owing to the continued error in her compasses,* she ran on a bank in a heavy gale of wind, and struck violently for some time. When she got free from the bank she was found to be very leaky, owing to her wooden false-keel, which had been carried away, leaving one of the bolt-holes open to the influx of the water.† This, however, was easily repaired, and she was found to be otherwise uninjured. On her return from the Baltic, with a cargo of timber, she was not so fortunate; but the circumstances of her loss were such as to show the superior endurance of iron vessels under the severest trials. In running down the North Channel, near Yarmouth, she was driven, in a heavy sea, upon the Cross Bank, the water on which was but eight-and-a-half feet, while the vessel drew thirteen feet. She got over it; and I have letters from several individuals, as well as the assertion of the owner, the captain, and the mate, attesting the extraordinary strength of the ship, and concurring in opinion that no timber-built vessel could have sustained the same beating, and that had she been of wood the crew and cargo must inevitably have been lost. When she had crossed the bank she leaked so much that it was considered prudent to run her ashore; and, owing to the awkward position in which she lay, and the small rise of tide, she could not be got off. After taking the ground she listed over to seaward and exposed her deck to the fury of the waves. This, I am informed, first gave way and

---

* The cause of this will hereafter be explained, and I beg to draw particular attention to the facts.

† Any false work under water, especially of wood, should be avoided.

left the sides of the vessel without support. One of the sides, at length, from continued working, gave way also, and rendered it impossible to save her. Her loss was therefore to be attributed to one of those casualties to which vessels are sometimes exposed, and which not even the strongest ship can withstand.

The "Royal George," one of the iron steamers running between this port and Glasgow—a vessel of unusual length in proportion to her beam—when loaded with about 150 tons of dead weight, besides her engines and coals, got on a rock, near Greenock, at high water, and was left there during a tide, without sustaining any injury. She rested nearly on her centre, and all who saw her were of opinion that no timber vessel could have remained in that position without breaking her back.

A large casting, weighing, I am informed, about four tons, fell into the hold of the "Princess Royal," Glasgow iron steamer, in consequence of the tackle giving way; but without causing any damage to the vessel. It was thought that no wooden vessel ever built could have withstood this shock.

Mr. Foulks, master of the "Alice," a small iron steam vessel built by us, and running hence to Runcorn, states that he has been for eighteen years mate and master of steamers, but had never met with her equal, especially for taking the ground, which she is obliged to do every tide. The engineer gives the same opinion of her adaptation for that service.

Captain Chaplin, who has had upwards of twenty years experience in steam navigation, and who was for some time manager of Woodside Ferry, in the course of some remarks on the strength of iron vessels, says:—"I may give you a case in point. The "Cleveland," built by you, got ashore amongst the rocks, in an ebb tide, where she was left high and dry for seven hours, hanging entirely by the heel and forefoot, without sustaining injury either in the hull or engine."

The "Nun," iron steamer, built by Mr. Laird, belonging to the same company, was lately placed in a most hazardous position, one end of her having caught on the pier, and the other settled down, on the retiring of the tide, on the adjacent bank, leaving the whole of the centre (a length of $81\frac{1}{2}$ feet, and containing the whole weight of the engine and boilers) entirely unsupported. She remained some hours, without injury, thus resting on two points, exposed to a trial of strength that would have ruined the strongest wooden vessel.

The power of iron vessels to withstand the action of the waves may be estimated by the fact, that when timber-built steamers have been wrecked, the hull and machinery have all broken up, while the iron boilers alone remained to indicate the spot where the disaster occurred.

At the present time we are at work on the iron steamer belonging to the Egremont Ferry Company, and are giving increased strength in her floorings. The bottom plates present a remarkable instance of the strength of iron vessels. When at full speed, she struck on the timber head of a sunken vessel, and, although the blow was most violent, the only injury sustained was an indenture of the plates, unaccompanied by any fracture or leakage. In repairing her, we have not thought it necessary to remove the plates, they being perfectly sound.

STOWAGE.

The stowage is the next question for consideration, and, though of great importance, it is less difficult to explain than the former subject, being simply a question of figures. The shell of a timber-built vessel is so much thicker than that of an iron vessel, that, with the same outside dimensions, the latter is frequently 18 inches wider and 12 inches deeper in the hold than the former. Taking the most favourable part of a vessel—namely, in the centre of the length—in a vessel of 200 tons, the proportion in favour of the iron vessel will be as 5 to 6; but in the ends, which are drawn finely off, the disparity is much increased, making the proportion of the whole contents about as 5 to 4. Supposing, therefore, that a vessel built of timber could stow 200 tons, she would, if made of iron, have room for 250 tons. The total capacities of the largest vessels will probably not approximate nearer than as 5 to 6; making the iron ship of 600 tons not to exceed in outward dimensions the timber one of 500 tons. The advantage of this is very great, and enables an iron vessel to trade, and remunerate the owners, in cases in which a wooden vessel would not return a profit; for, if we suppose that the freight of a 500-ton ship would just pay the expenses required to navigate her, an iron vessel would leave the freight on the extra 100 tons as clear profit.

With steamers, the comparison is much more in favour of iron than with wooden vessels. As the average cargo of a steamer is only, perhaps, about one-third of the load, in engine, cargo and coals together, an increase of one-fourth in her stowage adds 75 per cent. to her capacity for carrying cargo. In other words, to carry a given cargo, the iron steamer may be much less than a wooden one.

SAFETY.

The safety of a vessel is, in a great measure, identical with its strength; and the cases I have adduced with reference to this point as regards iron vessels, will be sufficient, I trust, to satisfy all reasonable minds. In addition, however, to their extraordinary strength, iron vessels afford protection both to life and property against the most awful accident that can befal a ship at sea,—namely, against fire. I admit that it can signify little to unfortunate passengers of what material the hull of a vessel is made, if her cargo, deck, cabins, and masts are consumed; as no one, who might escape the conflagration, could then remain in her. But, with ordinary precautions, it would be nearly impossible for a fire to take place, or to gain head, in the hold of an iron ship, provided the hatches were properly secured; for the bulkheads make each division perfectly air-tight, and effectually stop out the atmospheric air, without which fire will not burn,—thus confining the injury, when it does occur, to the compartment in which it originates.

Iron vessels are free, too, from the risk of starting the buts, or opening their seams, in a long continuance of heavy weather. They are much stiffer under canvas than timber vessels, and all who have sailed in them speak of their extraordinary buoyancy,—qualities tending very much to increase their safety. Their lightness of hull above the water, and their diminished depth, render them less crank than other vessels, and their buoyancy may be attributed to the same cause. Mr. Jackson informs me that the "Ironside" has never damaged her cargo, and stows without dunnage.

SPEED.

Shipowners are now becoming sensible of the value of fast-sailing ships; and I trust that a better system will supersede that which has hitherto guided them in the choice of models. The late register laws, which caused more injury to our mercantile marine than those at present in operation are ever likely to repair, and which originated and fostered evils that it may require ages to correct, are no longer obstacles to improvement, and their abrogation affords us some hopes that we shall not for ever be left behind other nations in the speed of our vessels. I believe I may safely assert that when iron becomes generally used for the purpose of ship-building, a much finer form may be adopted, without incurring the disadvantages that arise under the same circumstances, in even the most approved construction of wooden vessels. The material being much lighter to attain the same strength, and occupying less space, the model of the vessel may be made finer, and better adapted for high speed, without a corresponding loss in the amount of stowage and carrying qualities; and iron vessels, as they have more buoyancy, are not so liable as timber-built vessels to pitch when in a heavy sea. These assertions are not the result of mere theoretical speculation, but are derived from a long course of observations, guided by the opinions of many practical men who have become converts to the adoption of iron as a preferable material to wood in the construction of vessels.

DURABILITY.

It will be unnecessary to extend this portion of our inquiry by considering the progress of decay in iron vessels when employed in fresh water; but to direct our attention to the effect produced upon them by salt water, as a favourable result in the former case will not be disputed, if it can be established in the latter.

It is, as yet, impossible to assign any period for the duration of iron vessels in salt water, inasmuch as they have not been tried for a sufficient time to enable us to ascertain this point with precision.

The want of this proof must, however, be considered favourable, from the fact that were their decay as rapid as that of wooden vessels, such a result would already have become manifest.

The instance of the "Aaron Manby" will be in the recollection of the reader. She was launched in 1822; has been alternately employed in fresh and salt water, and is stated to be yet in good condition. The "Garryowen," has been about eight years in salt water, and is not yet perceptibly injured by decay. But the iron vessels of the Clyde probably furnish us with more decided proofs than any others of the durability of the material in salt water. It is well known that some of the first built of those vessels were extremely slight; yet it is truly surprising, when the frequent rubbing on the banks to which they are exposed is considered, how little effect time seems to produce upon them. So slight, indeed, is the apparent decay when the vessel is in use, and so much slower is its progress than is exhibited by iron when applied to other purposes in salt water, that many who have observed the fact are led to suppose that some occult and preservative law is in operation peculiar to iron so employed. A similar effect is said to be observable on the iron rails of a railway, the corrosion of which appears to be much less rapid when they are acted upon by carriages passing along them, than when they are lying in detached bars on the road-side.

The first iron steamer on the Clyde was, I believe, the "Fairy Queen," launched about twelve years ago, and not long since employed on this river. She was originally made of very light plates; subsequently underwent several alterations; and, I am informed, had her *third* set of engines on board when she was condemned. Shortly before she was broken up, and while she lay dry on the shore, we drilled some holes through the original plates, and found them to be tough and strong, with very little loss from corrosion. It cannot be said that the durability of these vessels is owing to any extraordinary care in coating them; for, after new and faster boats superseded them in the peculiar trade in which they were at first employed, little attention was paid to them in this respect, and it is not unusual for them to remain for three or four years without being cleaned or coated.

We have several favourable examples of the durability of iron vessels in the estuary of the Mersey,—particularly in the Woodside ferry boats, which have been in operation for about six years. I was engaged in building both the hull and engine of the first of these that was set to work,—the "Cleveland." The "Eliza Price," made by Mr. Laird, appeared immediately afterwards.* About twelve months ago we took the boiler out of the "Cleveland." During the time she had been at work—five years, and always in salt water—the parts of the shell under and about the boiler had never been coated or even seen.† Several persons observed, and minutely examined, the appearance of the plates in these parts. At the bottom, and up to the water line, there was not the slightest perceptible decay. A little above this, a slight scale was seen, which increased in thickness as it approximated the deck. This was caused by the leakage of the deck falling on the rounded part of the sponsons, where the plates were constantly kept warm by the boiler, and had not the advantage of the cooling effect of the external water. A coat of tar, or common paint, once a year, would have effectually prevented this slight corrosion; but, from the appearance of the parts constantly immersed in water, one would be led to suppose that coating in that quarter is superfluous.

As oxidation is promoted by heat, the boiler should be kept as far as possible from those parts of the vessel which are above water, or means should be adopted to keep them cool by interposing non-conducting substances. The plates in the engine-room do not appear to be more liable to corrosion than the rest, thus exhibiting a result different from that which was expected. It may be here stated, that it is perhaps not well to connect the copper pipes with the shell; and, to avoid this, we have latterly, in completing iron steamers, made short lengths of wrought-iron pipe, with flanches, and rivetted them to the vessel.

The absence of motion in the joints and seams, arising from

* Some well-known ship-builders of this town, when these boats first appeared, told a friend of mine that they would allow them a duration of three years, by which time, they were of opinion, they would certainly be eaten through, especially from internal corrosion. I trust they are now convinced that they were mistaken in their prognostication.

† At a late meeting of the Dock Committee, at which the propriety of having an iron light-ship was discussed, and a majority of 10 to 6 was obtained in favour of iron, the merits of these vessels were brought under consideration; and some of the opponents of iron as a material for ship-building stated their sound condition was to be accounted for by their having been coated four or five times a year!

strains, may be considered an important security against wear and tear. The durability of a substantial iron ship, always water-borne, will, I think, depend (accidents alone excepted) on the question of corrosion; and this again, in a great measure, on the care that is taken in cleaning and painting, which, when duly exercised, will be found to be of much advantage. But we have another great source of encouragement as regards the durability of iron vessels. We are now reasoning from cases where little or no attention has been devoted in the application of anti-corrosive substances—cases in which not only extremely thin plates have been used, but in which the greatest neglect has frequently been evinced. Have we not every reason to expect great and beneficial results in the preservation of the material, from the labours of the chemist, and the aid of theoretical investigation? Amongst the gentlemen engaged in this pursuit, Mr. Mallett, Civil Engineer, of Dublin, is worthy of the highest commendation; and it may be confidently anticipated that we shall, ere long, be put in possession of the means of rendering iron vessels much more durable than they are at present. Mr. Mallett, in describing the results of his experiments on the corrosion of iron in salt water, made at the instance of the British Association, and in allusion to his patent anti-corrosion process, supposes it quite possible in the progress of the art, to render iron as much proof against oxidation as "gold or platinum." Under the disadvantages, however, of the comparatively novel application of the material in lieu of wood, the loss from corrosion is by no means of an amount calculated to cause the merchant or shipowner to hesitate "which of the two to choose."

We now turn to the consideration of the durability of timber-built ships. And where, I would ask, in the catalogue of objections, real or fancied, to iron ships, is there one to be found equal to that dreadful scourge to wooden vessels—the *dry rot*; the effects of which are too well understood by shipowners to require any lengthened remarks from me? I should not, however, do justice to my subject did I pass it over in silence. No age has been without its nostrums, its quackeries, and its "infallible remedies" for the *dry rot,* and no period has been so productive of them as that in which we live; but, from all I can perceive, this plague is as prevalent as ever. A circumstance which has recently fallen within my observation is strongly illustrative of this subject, as involving the comparative merits of wood and iron. On removing the timber-work of the "John Garrow," preparatory to the alterations determined upon in that ship, we found that the dry rot had already (though she was not three years old) begun its work of destruction. That part of the lower deck which had been laid some time after she was launched had become decayed, and all the timber not exposed to the air was more or less affected. The iron, on the contrary, though it had been exposed to a fourteen-months' voyage, without being either cleaned or painted, was in a most satisfactory state. Not only were the plates and frames free from any perceptible injury, but the edges of a number of square bolt-heads, in and about the rudder, still retained their sharpness and appeared to be perfectly free from corrosion. The dry rot in wooden ships (which finds no parallel evil in those of iron) is frequently as remarkable in the earliness of its commencement, as it is invariably rapid in its progress; and no appliance hitherto resorted to has, in all instances, been effectual to avert its insidious development, or to arrest its destructive progress. The tree which is intended for a timber ship is no sooner felled than the oxygen, which, during vegetation, is held in harmless combination, begins its work of decay. The gradual combustion and deterioration of the woody fibre supervenes; carbonic acid is evolved; and the wood, becoming carbonized, loses its flexibility and strength. The progress of this decay is sure, though, by care, its effects may be retarded; but, from the numerous instances of its rapid increase, even in well-built vessels, we are led to conclude that no human foresight can prevent its poison from spreading, like a baleful epidemic, into every portion of the ship. How many stately vessels are now mouldering away under this destructive visitation, while their fine and graceful forms conceal the treacherous enemy within!

Some kinds of timber are much more liable to this defect than others, and its progress is much increased by neglect and a want of air. But when it has once commenced in a ship, her character for seaworthiness is lost, and she can no longer be depended upon; for it is too often found that she is weakened in the most vital points; that these parts, on which her safety mainly depends, are decayed; and that some unusual and casual strain is alone required to complete her destruction.

The action of iron fastenings is sometimes found to be as

injurious to the wood as that of the wood to the iron; and thus one of the main sources of support to a timber-built ship not only accelerates her decay, but is, in its turn, destroyed with equal rapidity by the wood. Figure 1, Plate 6, represents an iron bolt of an inch in diameter, which secured an iron knee to the oak rib of a well-known steamer in this port. The sound part was in the knee, and the decayed part in the wood. It was in its place only three years; was much above the water-line, and not near the copper; but was probably destroyed by the acid of the oak. This is a striking proof of the insecurity of iron fastenings when immediately attached to wood; but where iron is in contact with iron, the corrosion seems to be comparatively slow. There is also another point connected with the comparative durability of the two materials, which is very favourable to iron. Oxidation proceeds no faster on the surfaee of thick plates than on that of thin plates; so that plates of half an inch have double the durability of those of only a quarter of an inch in thickness. A large and consequently valuable vessel, if built with plates of proportionate thickness, will therefore be durable in increasing proportion to her size. On the other hand, I am of opinion that the thickness of wood, either as applied to planking or framing, does not increase its durability in a like proportion, but that, from the difficulty of thoroughly seasoning it when it is of great thickness, it is rather detrimental, and renders the small vessel as durable, or even more so, than the large vessel. I need not press this part of the subject further, nor descend to a comparison between iron ships and that extensive class of vessels now crowding our docks, to the serious injury of our own shipbuilders. I allude to a great proportion of those built in our North American colonies and sent here to be fastened with iron knees. The regulations at Lloyd's clearly show in how little estimation these vessels are held in point of strength and durability.

REPAIRS.

The wear and tear of iron vessels (and I speak confidently from actual experience) are practically trifling, and the repairs are consequently light. This item, which in wooden vessels presses so heavily on the profits, is, in iron vessels, of but slight importance; and although the comparison will be found very favourable in iron sailing ships, the fact will be more clearly shown by reference to steamers. The usual calculation for a timber-built steamer is, that the expense of repairs will in ten or twelve years, have equalled the first cost. In a well-built iron steamer repairs will not, I believe, have become necessary within that period, provided the vessel has not been injured by accidents ; and, under any circumstances, I feel confident that it will be more expensive to keep in repair the copper sheathing alone of a wooden vessel, than to effect the whole repairs in the hull of an iron vessel.

I revert to canal boats as furnishing a fair demonstration of the comparative merits of wood and iron. My informant in this point states that iron vessels are kept in repair at very little expense, but that wooden boats, when four or five years old, begin to be very expensive ; as much so, indeed, in one year, as iron boats will in three or four, or even more. He mentions boats which he knows to be ten or fifteen years old respectively, and which are still in excellent condition. "I know of some," he says, "that have been at work upwards of eleven years, and have not cost more than £2 each during all that period." Messrs. Tod and M'Gregor, of Glasgow, in writing on this subject, say,—"Some of the vessels we built were not in dock, or on a slip for four years : their bottoms were never seen, yet when examined they were found to be as entire as on the day they were launched. The hulls of some of our iron river steamers have not cost £1 each these five years. Knowing their stability, we are very anxious to get sea-going vessels of iron introduced. The 'Royal Sovereign' was, we believe, the first regular sea-going iron vessel in this country. The first six months she ran 25,000 miles under great disadvantages, she having to encounter dark nights and low water in coming up the Clyde ; yet during this time lost not one trip, but was punctual to the hour advertised."

In the event of accidents, the repairs of iron ships are extremely light, and in this respect also they bear a most favourable comparison with wooden vessels. The cases that have fallen within my own observation are so numerous, and so decisive of this fact, that, were it not superfluous, many pages might be filled in recording them.

The "Nemesis," war-steamer, ran on one of the Scilly rocks

in the British Channel with such violence that, to use the expression of my friend Mr. Claxton, the harbour master of Bristol, "if she had been of wood she would most probably have left her bones there." The damage she sustained by the shock was, however, so trifling, that she was navigated round to Plymouth, where the repairs were easily effected for about £30.

It is a great mistake to suppose that iron vessels are more difficult to repair than wooden vessels; and, in pursuance of this error, it is often urged, as an objection to them, that, should accidents occur to them while in a foreign country, no one could be found capable of making good the damage. But so far is this from being the case, that it is found that nothing is easier of accomplishment; for the injuries are generally external, and confined to a single spot, and may in most cases be sufficiently repaired by any ingenious man, if the ship be but provided with a few drills, some spare plates, bolts, and other necessary articles, which no captain should neglect to carry amongst his stores. Besides, when iron vessels become more generally employed, the necessary assistance and means will be found at all considerable ports and stations at which they may touch. But, on the other hand, take the case of a wooden ship, after she has received shocks such as were sustained by the "Nemesis, the "Iron Duke," or the "Vulcan," (if, indeed, she ever again floated;)—her whole frame dislocated,—leaking at every seam,—several floor-timbers broken,—the crew fearful at every moment that her continued working will increase the influx of water beyond the power of the pumps to keep it down;—she is run ashore at the first convenient spot, to avoid a greater disaster,—some temporary repairs are effected at much trouble and expense,—she returns home at great risk,—goes into the graving dock for one or two months,—her copper and planks are necessarily stripped off,—the broken flooring is at great cost replaced,—new planks and copper are supplied,—and, finally, the whole being caulked, she is at length ready again to put to sea! There is no ship-owner of any experience but will readily admit the correctness of this statement, and bear ample testimony to the expense incurred by such accidents,—an expense, indeed, frequently so far exceeding all previous calculation that the building of a new ship would, in the long run, be more in accordance with economy and discretion.

COST.

The first cost of iron vessels is next to be considered. The public, and frequently builders themselves, are under considerable misapprehension in respect to the comparative expense of wooden and iron vessels. It is the general impression that iron vessels may be built at a much less expense than wooden ones; and some builders have, consequently, been injudiciously induced to take contracts at estimates too low to ensure them a remuneration for the use of adequate strength of material, and for fidelity of workmanship. On this point, I would strongly counsel ship-owners, in making contracts, not to pursue this system of mistaken economy, the result of which may be readily foreseen; but by a wiser liberality to secure the more perfect and satisfactory accomplishment of the object they have in view. No one, it may be here also remarked, can avoid observing and lamenting the low state, in this country, of what may be termed mercantile naval architecture, in which men of science meet with little or no encouragement to attempt improvements, and have become weary of a system which for many years has brought them no return. But let us hope for better things in iron ship-building; let us trust that both owners and builders will see that their interest lies—the former in procuring good sound vessels, and the latter in obtaining a price that will leave no excuse for imperfect work!

From a careful consideration of the question of cost, I have arrived at the following conclusions:—

1st. That a good serviceable iron sailing vessel, not exceeding 300 tons burthen, will be equal in cost to an English-built twelve years timber-built ship, of the same external dimensions, without including the price of copper sheathing for the latter.

2d. That iron vessels of above 300 tons have the advantage of being rather less expensive than wooden vessels.

3d. That for very large merchant vessels, and ships of war, iron will also be found to be much less expensive than wood.

4th. That iron sea-going steamers bear also about the same proportion according to their different sizes.

5th. That iron vessels, for rivers, may be built at a light expense, but so built are unfit for sea service.

This estimate is to be considered to apply only generally, as the cost of either wood or iron vessels is ruled by the cost of the materials, and the style of finishing.

It must be always borne in mind that builders of iron vessels cannot have the advantage, in one point, which is enjoyed by ship carpenters: they can never derive much profit from that which to the latter constitutes the cream of their business, namely, the *repairs*. What moderate profit can be realized, in these days of competition, must be derived mainly from the building of new vessels.

Viewing the subject in another light, we perceive other sources from which the public are to be gainers by the introduction of iron vessels. Under the head of "stowage," I have shown the advantage iron presents over wood in respect to the increased internal capacity which ships constructed of the former possess over those built of the latter material. I have stated this to be in the proportion of 4 to 5 in small, and of 5 to 6 in large vessels. An iron vessel required to carry 250 tons, therefore, equals, in the first cost, a wooden vessel that will carry only 200 tons: and an iron vessel required to carry 600 tons will cost less than a wooden vessel built to carry only 500 tons. I have before shown the advantage possessed by iron for steamers as respects stowage. The same reasoning also applies to the first cost.

DRAFT OF WATER.

The remaining point in which I have to prove the superiority of iron over wooden vessels is that of light draft of water. Iron sailing vessels may be built of any requisite depth and sharpness for holding on in a sea-way; but, where a light draft is essential for a peculiar service, it may be attained to a greater extent by the use of that metal than by timber. This advantage, of course, arises from the weight of iron necessary in the construction of a vessel being much less than the weight of wood required for the same purpose. When iron vessels of very light draft are employed, means should be adopted—and many may be resorted to—to prevent them from drifting to leeward. The lee-boards of Dutch vessels, navigating shallow waters, and the sliding keels in American pleasure-boats and other vessels, are very efficient in this respect.

Iron vessels admit of the latter plan with great readiness and security; and, when adopted, it has been proved to be highly successful.

THE "JOHN GARROW."

Some particulars of the "John Garrow" may be here introduced, as bearing on some of the questions under discussion. She was built in Aberdeen, and launched in the early part of 1840. Her burthen, by the old system, was 555 tons; and, by the new measurement, 711 tons. She set sail for Bombay, with a very heavy cargo, in April of the same year, and was absent upwards of twelve months. After her return, she remained in the docks until we commenced the alterations in the hull, now in progress. In consequence of her extremely full and round build, particularly abaft, she steered imperfectly, and sailed but slowly when deeply laden. She consequently made a tedious outward voyage. At Bombay she took in a considerable cargo; but, having been a long time in harbour, her bottom became foul, and no means being taken to clean her, her homeward passage was also prolonged. Some damage accrued to the cargo, which I shall explain hereafter; and under the section of my subject relating to fouling, I shall describe the condition of the hull, which, on being cleaned, was closely examined by several practical and scientific men, and found to be in a very satisfactory state.

It is a matter of no surprise to find that the unfortunate results stated were eagerly seized upon by the opponents of iron vessels and held up by them as proofs of their inefficiency and failure. The largest shells that had adhered to her bottom were selected and carefully preserved as trophies against those who advocated the cause; but it is due to the enterprising owners of the vessel to say that they were not for a moment disheartened; and I am confident the day is not far distant when we shall feel that we owe much to the discretion and firmness of these gentlemen; for, had this vessel been the property of men of less foresight and perseverance, she would probably have been condemned. Such a decision would have operated as a severe blow to the progress of building iron vessels, from which a long time would have elapsed before it could rally, though the principle would ultimately have worked out its complete vindication. Fortunately, Messrs. Anderson, Garrow, and Co. were satisfied that, in spite of the drawbacks referred to, there was nothing to despair of; but, on the contrary, much to inspire confidence in the success of iron vessels, and to urge on to such improvements as would achieve it.

The "John Garrow" had been at sea, without any examination of her bottom, for upwards of twelve months. She had (as stated) carried large cargoes to and from Bombay. Her iron framing and beams were extremely slight, and the timber-work was imperfect: yet, when strictly examined on her return, not the smallest appearance of strain or decay was observable in the iron shell, and when cleaned and scraped, there was nothing in the iron work to indicate that she had ever been subjected to any severe test. In alluding to the defects of this vessel I shall simply state what I myself witnessed. How these defects arose, it is not my province to inquire; but I consider it essential, in order to vindicate the principle of iron ship-building, that they should not pass unnoticed.

I need say but little of the general appearance of the "John Garrow." She was seen by thousands in this port, and every one agreed, I think, that she could not possibly sail well when loaded; an unusually bluff bow, full quarters, and low transom—all militating against her speed, and not less so against her steering. Her rudder was also on a bad principle. It was both small and thin, and, being attached to a broad stern-post, half of it was probably useless. The iron framing was very light in proportion to what has been adopted in other sailing vessels, each frame consisting of only single three-inch angle-iron, and about two feet six inches asunder. The beams were formed by two pieces of angle-iron, rivetted together without any plates to stiffen them. The stancheons were of round bars, of about an inch and a half only in diameter, except where strengthened by angle-iron. These and the lower-deck beams were so slight as to be bent in all directions by the mere pressure of the cargo. The plates, however, were unusually strong, and it was this circumstance, together with the test she had been put to, that encouraged the owners to have her altered and amended. But I would ask any unprejudiced man, whether, in all this, any sound argument can be brought forward against the use of iron as a material for ship-building?

This vessel, almost without internal support, and entirely without ceiling, except what covered the bottom, encountered a great deal of heavy weather, carrying a large cargo of dead weight, and all without an apparent strain of any kind whatever.

I have said that the cargo, on her homeward passage, was damaged by the water entering her sides and deck. This was supposed to arise from leaky joints. Such, however, was not the fact. It arose from the following circumstance. The wood-work was in so defective a state that, in the contemplated alterations, the whole was condemned. In removing it, we found an omission, which was one of the principal causes of the damage done to the cargo. A broad African oak ribbon had been placed round her, outside, covering about two feet in depth of the iron plating. To secure this, several holes had been made in the iron plates; but, as some of these holes were partially or entirely concealed by the iron ribs and stancheons, all of them were not discovered, and they were consequently left open. From the heat of the sun, and other causes, the wooden ribbon sprang off from the sides of the vessel, leaving nothing to prevent the water from pouring in by the holes every time the gunnel, by a lurch or a rising sea, was under water!

I notice these facts for no other purpose than to show that the defects of the "John Garrow" were not attributable to her being of iron; and it would be unjust to adduce her misfortunes as arguments against the principle, if properly carried out, of building ships of that material. So little, indeed, were her defects to be ascribed to this cause, that, although she was lamentably deficient in internal support, not a leak, nor the appearance of a strain, could be observed in the iron work, while the *timber-work* was in a deplorable condition. The owners, with a view to remedy the several evils referred to, had an accurate model of her made while she was in the graving dock; and they subsequently gave orders for considerable alterations and improvements in her hull. The principal of these was, that she should be lengthened by about 14 feet, which enabled us to entirely remodel both ends,—reducing the heavy quarters, and raising the main transom about 4 feet;—and, also, in place of her original bluff bows, giving her a comparatively fine entrance.

In effecting these improvements, now in progress, we have

adopted all the most approved principles that our own experience and that of others conversant with the subject could suggest. The stem and stern posts are formed of large bars, or rather strong flatted pieces of solid iron, the heels of which run several feet along the bottom of the keel. The form of the main body remains unaltered, as no improvement was here required. A new and strong iron keel, eighteen inches deep, has been attached. The iron frames have been doubled throughout, by rivetting to them bars of angle-iron in a reverse position, leaving a flat surface on which to bolt a wooden ceiling, in the manner before described. Great additional strength has been given to the floorings, and the upper and main decks have been entirely renewed. These last are worthy of some description. They are each composed of three pieces of angle iron, rivetted together. The two upper pieces are three inches in each direction, and between them is placed a piece six inches by two-and-a-half inches, the two-and-a-half-inch horizontal surface forming the under part of the beam. This form of beam is adapted to prevent injury to the cargo caused by the sharp edge of the under part of the beams, as usually constructed. Figures 1, 2, 3, 4, and 5, Plate 2, show the upper and lower deck beams on this improved principle, and also the mode of attaching them to the sides. The upper bars are bent at right angles, and carried down the sides a short distance; and the under piece constitutes a knee, by being bent downwards, as shown in the drawing. On the upper sides of these beams, flat bars of iron are placed—two to each deck—running the entire length of the vessel, as shown by *a a*, Plate 5. These are firmly rivetted to all the beams, at the point of crossing, and are covered by the deck planks. To the lower deck, a stringer of the strongest angle-iron has been added, rivetted both to the frames and beams. *(See Fig. 1 and 2, Plate 2; and Plate 5, at b.)* The old stancheons have been all removed, and round bars, of three inches and two-and-a-half inches in diameter, substituted. The sides of the vessel have been let out in midships six inches on each side, and the sheer of the deck has been much improved.

These few particulars are valuable as showing with what facility iron vessels may be altered and strengthened at pleasure, and under circumstances that, in wooden vessels, would render such operations almost impossible. All the additions that have been made are as strong and efficient as if they had formed part of her original construction, and the vessel bears no appearance of patching. Many of the plates and angle bars that were removed have been restored, and no one could now discover any difference between the old and the new parts. Not so with the wood work: scarcely any of it could again be made use of.

I have added to this communication a draft of the "John Garrow," as she was originally built, and another showing her present model. She will not only very much exceed all other vessels of her size, now afloat, in her capability for stowage, but, it is expected, will be greatly improved in speed.

THE "GREAT BRITAIN."

The scientific world is awaiting with intense interest the completion of the iron ship the "Great Britain," now building at Bristol, and better known as the "Mammoth." I have been favoured by Mr. Guppy, the Managing Director of the Company to which she belongs, with her dimensions, which are truly gigantic. They are as follows:—

Length of keel	289	feet.
Length from figure head to tafrail	320	,,
Beam	51	,,
Total depth from underside of the upper deck to the keel	31	,, 4 inches
Draft of water when loaded	16	,,
Tonnage, per old measurement, about	3500	tons.
Displacement of water when drawing 16 feet, about	3000	,,

The plates of the keel are from $\frac{3}{4}$ths of an inch thick in the middle, to 1 inch at the ends; and all the plates under water are from $\frac{5}{8}$ths to $\frac{1}{2}$inch, at the top, except the upper plate, which is $\frac{5}{8}$ths. She is clencher-built, and double-rivetted throughout. Towards the extremities, and quite aloft, the thicknesses are reduced gradually to 7-16ths.

The ribs are framed of angle-iron, 6 inches by $3\frac{1}{2}$, by $\frac{1}{2}$-inch thick at the bottom of the vessel, and 7-16ths thick at the top. The mean distance of the ribs from centre to centre is 14 inches; and all these ribs will be doubled: the distance is then increased to 18 inches, and then gradually to 21 inches at the extremities.

The boiler platform is of plate iron, supported upon ten iron kelsons, of which the centre ones are 3 feet 3 inches deep. These kelsons are formed, like the floorings, of iron plates placed on edge.

The hull is divided into five distinct compartments by means of substantial water-tight iron bulkheads.

The decks which are of wood, consist of the cargo deck, two cabin decks, and the upper deck.

It would be an endless task, and, without the aid of drawings, a fruitless one, to attempt a description in detail of the construction of this magnificent vessel. I can, however, state that her lines are very beautiful, and adapted to attain the highest rate of speed. The general character of the workmanship is very good, and does great credit to the builder. Her exquisite proportions prevent that appearance of heaviness which is generally observable in large ships; and, although she is probably the strongest vessel ever built, she has a remarkable air of grace and lightness.

OBSTRUCTIONS TO THE PROGRESS OF IRON SHIP-BUILDING.

It will now probably be asked—"Why have not iron vessels become more general, if there be any truth in the opinions and assertions here brought forward; for, if such advantages really existed, would they not long since have become manifest to the most superficial observer?" I will not evade this enquiry. On the contrary, I gladly enter upon it, as I hope to explain the principal obstructions which have presented themselves to the progress of ship-building in iron; and to satisfy those who would investigate the subject dispassionately that the usual objections to the employment of that material are without foundation.

THE COMPASS.

The assumed difficulty respecting the compass was a fair ground of objection. Iron vessels had long been used on canals and rivers before any one ventured to send them to sea. And as it is to the compass that we are indebted for the ease and safety with which vessels can be navigated when out of sight of land, it was natural that the derangement which it might suffer in an iron vessel should be viewed with alarm; and thus the objections of seamen, whose movements are so dependant on that delicate instrument the compass, were well grounded.

The enquiry with respect to the magnetic attraction of iron vessels is one of extreme interest, and well worthy the attention of scientific men. Professor Airy, Astronomer Royal, has identified his name with this subject; and others of the highest attainments have exerted themselves to investigate the peculiar laws by which the electric current is guided, and to discover methods to counteract its effects on compasses, especially those of iron vessels. It was ascertained by Captain Johnson, during some experiments in the "Garryowen" iron steamer, on the Shannon, that an iron ship becomes a large permanent magnet, having a strong magnetic current in the direction of the ship's length, and causing a serious disturbance to the general polar direction in which the needle points. I suppose, however, that this circumstance is favourable to its correction; for, as the disturbing force is nearly constant, means may be adopted to counteract or so to neutralize the disturbing current, and allow the needle to point in the same direction as it would were it removed from all local attraction. A deviation is found to exist in the compasses of most wooden vessels, arising from the pieces of iron employed about the deck. In ships of war the guns have a powerful effect upon the compasses, and many plans have been tried for their correction. The cause of disturbance is here, however, probably different from the principal part of that of ships built entirely of iron,—the latter, as I have before said, being in themselves permanent magnets; while, in wooden ships, the effect arises perhaps from the iron bolts, bars, straps, &c., in the neighbourhood of the compasses becoming conductors (though feeble ones) of magnetic currents which arise from other sources.

An iron vessel is formed of an immense number of plates and bars of different forms, having varying degrees of magnetic force, and the electric currents run in different directions. A curious question then arises as to the period at which the current begins to assume a uniform course through the vessel, neutralizing or turning into one channel the multitude of conflicting currents which pervade its body. I purpose, on the first convenient occasion, to make a series of daily observations with a compass on an iron vessel,—to commence with the first plate, to be continued till she is launched, and afterwards to compare these results with the deviations that may be observed when the ship comes to be turned in different directions. I think such observations, placed in the hands of men who have made the subject their study, may be the means of eliciting some curious and interesting discoveries.

I had, in common, I believe, with others, long entertained the opinion that the course of the magnetic current was determined during the operation of rivetting, or that it added considerably to to its intensity—its force also being influenced by the polar direction of the ship's keel, while on the stocks. I was led, however, to doubt this effect from some observations I made about three years since on an iron steamer called the "Alice," built for the trustees of the Duke of Bridgewater's canal.

As iron vessels are generally built, it is very difficult to make correct observations to ascertain the effect of rivetting. This operation proceeds at short intervals, and in different parts, as the plates are applied; and it would be almost impossible, should any alteration be observed in the compass, to ascertain whether it was caused by the plating or by the rivetting. Not having any premises near the water, where the "Alice" was built, we first erected her at our boiler yard, and afterwards re-erected her on a vacant piece of land at the water's edge, from which she could be launched in the usual way. It appeared to me that this presented an opportunity to ascertain the effect produced on a vessel by rivetting. When, therefore, the plates had been put together at the water's edge, and before the rivetting was commenced, I obtained a very delicate needle, and made a few accurate observations as follows:—I drew a long line a-head of the vessel in the direction of the keel; and having selected a point at a considerable distance, where I could perceive no effect on the compass, I ascertained the direction in which the vessel pointed. I then measured 100 yards from the vessels stem, and there fixed the compass on the plane of the lower side of the keel. I could not here perceive any deviation, but, as I moved forwards towards the vessel, the disturbance became very considerable. I carefully noted down all these results, and also made an observation under each quarter of the vessel. I now waited till the vessel was entirely rivetted and nearly ready for launching, and then fixed the compass in the same places as before, and again observed the deviations. These I found to correspond most accurately with the first deviations, except in the last point, which was only one yard from the stem ; but here the original support on which the compass rested had been removed, and I could not make any fair comparison. I think it possible, however, that the great number and size of the rivets which had been added, since my first observations, to the lower part of the stern, would have an effect on the compass, independently of the hammering. I merely make these observations in the hope that they may prove useful to some one qualified to reason upon them ; and only suggest that as the polarity of the ship is not affected by the rivetting, but seems to be dependent on the quantity or quality of the iron, so we may with more confidence expect that time will have but little effect in increasing or diminishing the magnetic intensity. Captain Johnson further observed that the compass, placed a few feet above the deck, was not much attracted by the iron, and it is a common practice for the captains of iron ships to place a compass in the cross-trees at the topmast head, by which to ascertain the correctness of the one placed in the binnacle.

When the "Ironside" was ready for sea, Professor Airy came to Liverpool and made a series of observations upon her, with a view to the correction of the local attraction of the ship. This object he effected in the most satisfactory manner; and not the less so that the plan he adopted was exceedingly simple. Mr. Airy afterwards, with great liberality, published in the *United Service Journal* a list of rules, by observing which, any one might correct the compasses of iron vessels, without difficulty.* This system is

* The operation may be shortly described as follows :—Having fully completed the equipments of the ship, especially as regards the iron-work, and having determined the exact position in which the compass is to be placed—take a point on the deck exactly under it—through this point describe two lines on the deck, one parallel to the keel and the other at right angles with it. Provide two powerful magnets, about two feet long, and a small box, seven inches long by three wide, full of small iron chain, or small pieces of iron, laid in different directions. The vessel, being in a wet dock, should be firmly moored by four hawsers, and her head being made to point exactly to the magnetic north, as ascertained by a delicate azimuth compass on shore,

found to be so efficient, that when applied to iron vessels the compasses are generally more correct than those in vessels built of timber. Few ships are free from some local disturbing influence ; but it is not considered to be so great in wooden ships as to require correction, and the commander having discovered its extent, makes allowances for it in his calculations. The compasses in iron ships are, however, so accurate, after being corrected, as to be free from any sensible error.

The power of the magnets, and the intensity of the magnetism of the ship, will probably alter by time ; but I have not heard that this effect has yet been observed on any compasses that were well corrected. A careful navigator would, however, soon observe any deviation that might arise in them, and on his return to port have them again corrected ; as iron vessels become more general, some one will be found in each port capable of performing this operation with sufficient exactness for all practical purposes.

THE "IRON DUKE."

I now recur to the case of the "Iron Duke," the loss of which is attributed solely to the inaccuracy of the compasses, and her bad steering qualities—defects erroneously stated to be inseparable from iron vessels. The latter objection is too absurd to require an argument, as a timber vessel of the same form would have steered no better ; but the former deserves examination. On the return of the "Iron Duke" from India, the Captain reported that his compasses were not to be depended upon. The owner, in consequence, sent for Mr. Heron, of Glasgow, who regulates compasses on Professor Airy's plan, and they were by him re-adjusted. They were

---

an observation should be made. The needle will now probably be found to be very incorrect: one of the magnets should then be slipped along the athwart ship line, either over or under the deck, till the needle points correctly. The ship's head may then be turned due east, and the other magnet, being placed on the fore and aft line, is so regulated that the error, if any, in the needle, is corrected in this position also. The vessel should then be moved round to all the four points, north, south, east, and west, and any error now observed may be corrected by again changing the position of the magnets. It is now necessary to point the head of the ship towards the north-west, or south-east, and any deviation that is observed will be corrected by the use of the small box of chain, the exact spot for which must be determined by trial. These corrections carefully attended to will cause the compass to be free from all sensible error. The magnets should be of the best description, and be placed in a box full of tallow, which box may be nailed to the deck or the ceiling of the cabin.

again, however, found to be incorrect, and, it was ultimately alleged, were the cause of the loss of the vessel.

I had long felt satisfied that Professor Airy had found the true remedy; and as I had never before known his system to fail, I took considerable pains to ascertain the cause of the deficiency in this instance. I could not, however, discover anything satisfactory, and began to despair of being able to clear up the mystery,—when I was informed by the mate, in the course of conversation, that a variation of the compasses was invariably observable when the vessel tacked. On further enquiry, the mate informed me that the iron tiller projected towards the binnacle, and probably caused the deviation. With a view to obtain the most accurate information on this point, I wrote to the builders; and what was my disappointment on being told by them that the tiller was placed, as usual, *abaft* the rudder head. Meantime, the Report of Lloyd's Surveyor had spread throughout the kingdom: premiums of insurance on iron vessels advanced, and nothing was heard but doubts of their success, as no reliance, it was said, could be placed on the compasses. I did not however, yield to the general apprehension; and, on communicating with the Captain of the "Iron Duke," and making more minute enquiries, I became possessed of facts that set my mind at rest as to the cause of the derangement of the compasses of that vessel. The mate's account was in part correct. It appears that a large piece of iron, about fifteen inches long, projected from the rudder head towards the binnacle, provided for the purpose of applying a spare tiller, in the event of that abaft the rudder head giving way. From the Captain's account, I conjecture that this piece of iron reached within eighteen inches of the binnacle,—for he states that a man could just pass between them. I have since conversed with several persons who recollect having observed it; but, strange to say, no one ever thought of the probable consequences. Two circumstances perhaps checked enquiry;—one of them the supposed correction by Mr. Heron, which, being unsuccessful, no one examined further—imputing the deviation to the influence of the iron of which the hull was composed. If the same irregularity had occurred in a timber vessel—and it often does—every one would have looked round for a cause; but, in this case, not one of those who went down to examine the vessel at Yarmouth appear to have noticed the fact stated. So entirely, indeed, was it overlooked, that the spirited and intelligent owner of the "Iron Duke," Mr. Hartrick, of New Ross, who was much concerned and disappointed by the loss, and who also went to Yarmouth, was told of it for the first time by myself at Liverpool. He then recollected that orders had been sent to the builders to take unusual precautions to secure the rudder against accidents, as a vessel in which he was interested had been just then nearly lost by a breakage in her steering apparatus. Mr. Hartrick expressed a strong desire to have the subject fully investigated, and kindly promised me any assistance in the furtherance of the enquiry.

If, then, I am right in my opinion that this additional piece of iron, traversing in the immediate proximity of the binnacle of the "Iron Duke," was the cause of the derangement of the compass, is there not reason to complain of the Lloyd's surveyors and other persons sent to the spot expressly to trace the facts, and who wrote and published reports amounting almost to a condemnation of iron vessels?†

I can account for Mr. Heron not perceiving any disturbance from this piece of iron by supposing that the tiller remained fixed in one position (probably amidships) while he was correcting the compass. It is not sufficient to account for the accident by saying that the power of the magnets alters, or the compasses become dull, when surrounded by so much iron; or to suppose that the well-known variation observed in different latitudes does not follow the same course in iron as in wooden vessels. Almost from the first, the compasses in the "Iron Duke" were found to be wrong. The Captain told me that on his second voyage he observed them to be incorrect in the Irish Channel, which is navigated almost daily by iron steamers, the commanders of which report most favourably of the accuracy of their compasses. He also informed me that he had the very best compasses on board, but that all of them were affected in the binnacle; while, at the same time, there were parts of the deck

† A letter of this kind from Mr. Dalzell, of Glasgow, who saw the "Iron Duke" ashore at Yarmouth, was read by a gentleman at the Polytechnic Society. In this document Mr. D. gives a most favourable account of the strength of iron vessels; and of the "Iron Duke" in particular: but concludes by stating that in consequence of the incorrectness of the compass, he does not recommend them for long voyages,—although he observes that in the steam vessels it was always correct, and he does not appear to have noticed the error on any other occasion.

where all of them would act quite correctly. It is almost unnecessary to observe the fact of the great degree of polarity given to a piece of common malleable iron, from various causes ; and that the same piece of iron, under other circumstances, may have but little power to attract the needle. It is well known, from this fact, that one piece of iron may be nearly harmless, while another, to all appearance the same, may be a powerful magnet.

I now leave the subject to be judged of by the public. It is of the greatest importance that this circumstance should be accounted for ; and I have, therefore, dwelt upon it thus at length. To have declined noticing it, would have implied an inability to meet the objection founded upon it, and involve the greatest mischief to the cause. But I trust the explanation here supplied will have an opposite tendency, and induce shipowners to receive with more caution opinions hastily adopted against iron vessels.

I may add that I was told, by the captain of the "Iron Duke," a circumstance which, though in itself of comparatively trifling interest, may be deemed not unworthy of passing notice. A colony of rats took possession of the vessel previous to his first voyage, but, not relishing the obdurate material of which she was built, a portion of them deserted her at Calcutta, and the remainder abandoned her immediately on her return to England.

PREJUDICE AGAINST IRON VESSELS.

The question of the compass—which at first sight involved one of the most serious objections—being thus satisfactorily set at rest, it might be thought the use of iron in ship-building would have been patiently allowed to have a fair and impartial trial ; but, unhappily, to many persons connected with shipping, and whose ideas had long been wedded to wood, as the exclusively applicable material for that purpose, the novelty alone of the principle appeared in their minds to justify the vague opinions and surmises of its inefficiency, which, from the force of prejudice, were magnified into

‡ After writing these remarks upon the "Iron Duke," I requested Professor Airy to favour me with his opinion, which he did, in a letter, an extract from which will be found in the appendix, as well as his further opinion, in reply to my enquiries on several points connected with the compass in vessels, which will be found at the conclusion of his valuable communication.

insurmountable obstacles to its adoption. Its advocates were set down as visionaries, whose theories practice would speedily dissipate. Even men of considerable experience frequently asserted that *iron vessels would never answer*, and, as if anxious to hasten the confirmation of their opinions, spared no pains to search out or imagine objections, eagerly seizing upon every defect in those first constructed, and holding them up as satisfactory proofs of the failure of the principle, without giving themselves the trouble to think of the possibility of improvements. Time having already shown, at least to a great extent, that one of their first objections—the supposed rapid corrosion of iron in water—was no longer tenable, they began to lay great stress on the tendency which iron has to become foul, when long immersed in salt water; but I trust I shall hereafter be enabled to prove that this is to be attributed to neglect, and that the adhesion of weeds, shells, &c., is so far from being an insurmountable objection, that iron ships may be effectually cleaned, even while under sail. Besides, (as before stated,) chemical means will doubtless ere long be successfully employed for this purpose. The opposition thus shown to iron vessels is not only to be regarded as injurious, because retarding a scientific improvement, but as deterring shipowners and directors of steam companies from adopting the principle, and causing them still to invest large sums in the construction of timber vessels, which, before they are launched, have in reality lost one-fourth of their value by the rapid improvements in iron vessels.

Iron ship-builders themselves are not always free from blame, as having by their imprudence injured the cause, or impeded its progress by using material of inadequate strength, by carelessness in its scientific application, or by imperfect workmanship, frequently the results of accepting contracts that will not remunerate them for the construction of strong and well-built ships. In their eagerness to obtain business, they have allowed themselves to be guided and controlled by those who have orders to give. This should be firmly guarded against, as leading to the construction of vessels unfit for service, from their extreme lightness,—thus placing the principle of iron ships in jeopardy. Iron vessels, too, have generally been adapted for situations in which wood vessels have been objectionable or inapplicable, a light draft of water being the advantage sought; and it frequently happens that to obtain this every other advantage is sacrificed. To ensure great displacement, the models are generally too full. They will consequently neither sail fast (particularly on a wind) nor steer well, and they are sometimes almost without keels, or any other means of preventing them from drifting to leeward, a defect rendering them nearly unmanageable in a heavy sea. All these objections, which may be entirely obviated by the adoption of proper models, the employment of material of sufficient strength, and fidelity of workmanship, are seized upon by the opponents of iron vessels, and arrayed against them, as if they were inseparable from the principle. But the truth will yet prevail; for opposition, whether it arise from misconception, prejudice, or self-interest, though it may retard cannot prevent its ultimate development.

The objections are, however, becoming weaker; and it is amusing to trace the progress of these results on men's minds. It is first admitted that iron vessels are well adapted to fresh water; then that they may answer well as steamers, even in *salt* water; and thus by degrees every barrier to their adoption falls and gives way to the force, not unfrequently, of unwilling conviction.

Underwriters have not been the last to find objections and to seize every excuse to raise and keep up premiums, and hence originates a serious obstruction to the introduction of iron vessels. Improvements that increase the safety of ships are generally understood and openly avowed to be against the interest of underwriters, and it is not reasonable, therefore, to expect their support, even should they become convinced that great benefit is to be derived from the use of iron.

Many shipowners have been deterred from giving a preference to iron vessels from an apprehension that the public is prejudiced against them, and that passengers would object to sail in them. It is proved, however, that no such prejudice exists. The iron vessels running between this port and Scotland and Ireland are quite as popular as those of wood, performing the same voyages, when the speed and accommodations are not inferior.

FOULING.

The fact that iron vessels become coated in tropical climates with shells and weeds has operated considerably against them, but both the effect and the ultimate consequences are greatly exaggerated. Mr. Mallett, of Dublin, upon whose experience and extensive knowledge of chemistry much reliance may be placed, has given it as his opinion that chemical means will ultimately be adopted to remove this objection ; and several men of high chemical standing also concur with Mr. Mallett in this opinion.

The "John Garrow," in her voyage to the East, visited Bombay, and thence sailed direct to England. She was found on her return to be entirely covered by a dense mass of the shell called, by naturalists, the *balanus*, probably from its being found adhering to whales. There were very few of those found above a certain defined line, some feet below the low water mark ; but from that line to the keel they were in great quantities, excepting on the rivets, which were not covered. The "Iron Duke" shortly after arrived from Calcutta, but with this remarkable difference in her condition, that scarcely a shell of the description named could be found, and her bottom was very clean, having adhering to it only a small number of the animals called *otion*, and the common barnacle. The cause of this difference is easily explained. The "Iron Duke," while at Calcutta, was in fresh water, which always destroys the animals that are bred in salt water ; while the weeds which collect in the fresh water in tropical climates drop off as soon as the vessel reaches the sea. The "John Garrow" not having left the salt water, the shells which had collected at Bombay grew and increased in number on her voyage home ; and we may conclude, from the state of the "Iron Duke," that, had the "John Garrow" started from Bombay free from shells, they would not have collected on the voyage, and she would have come home with a clean bottom. The cause of the shells not adhering to the rivets, is probably to be traced to a slight galvanic shock which they receive from them, and it has been suggested that a galvanic battery might be employed, by which oxidation would be prevented, and the bottom be kept free from shells. These objects might to some extent be accomplished, but I fear the operation would be injurious to the compasses. Both the balanus, and the common barnacle are small and tender when they first adhere to the ship ; and there would be no difficulty in removing them when in that state, by means of a scraper properly contrived for that purpose. It is said that no difficulty is found in preserving the "Nemesis," and other iron vessels in the Indian seas from becoming foul; and I believe that a broom is sufficient to remove anything that is not allowed to adhere too long.

INSTRUMENT FOR CLEANING SHIPS' BOTTOMS.

After giving the subject considerable attention, and making some experiments, I constructed an instrument for cleaning the bottoms of vessels while afloat and sailing. The principle I adopted was to scrape the vessel in a transverse direction with a wooden scraper, drawn under the bottom by small ropes, worked by the men on deck. Annexed is a plan of the scraper, *Figures 2 & 3, Plate* 6, which will be readily understood. It is made of two pieces of ordinary deal timber, connected by two wooden bars across the back, and is about two feet long. To the ends of these bars, bridles are attached, from which lead two guy-ropes, *a, b*. There is also another guy-rope, *c*, leading to the fore-part of the vessel. Figure 4 is the section of a vessel, showing the scraper at *a*, under the bottom, with the ropes leading to it from both sides of the deck. Light brackets, with sheaves, *(see b, Fig. 4,)* are also provided, which are fitted to sockets prepared for them in the railing. The scraper must be adapted to the vessel on which it is worked; but, being extremely simple, a suitable one may be made by the ship's carpenter while on the voyage. The face of the scraper must be bevelled, so that in drawing it back, it may not catch against the ledges of the plates; and the depth from the hole through which the bridles *a, b*, pass to the edge of the scraper, must be rather more than the depth of the keel, so that when drawn close to it, the keel may not raise it from the garboard streak. The mode of working it is simple. It is drawn under the bottom, at the fore part of the ship; and the ropes, *a, b*, are led through the sheaves of the brackets. The operation then proceeds by the men alternately drawing the scraper down to the keel and up again. The rope *c*, which leads through a sheave near the bowsprit, is run out by degrees, and the way of the vessel through the water forces the scraper towards the stern, and it has a strong tendency, from the pressure of the water, aided by its natural buoyancy, to cling to the bottom of the vessel. By these means, the scraper also passes into the hollows of the fore and after parts of the ship.

As the rope *c* is run out, the brackets must be moved aft, a few feet at a time, so that the transverse course of the scraper is shifted six to twelve inches, and will traverse over every part of the vessel's bottom two or three times, or more, as may be considered necessary. Two scrapers may be used at the same time, one on each rope, *a* and *b*, the one being drawn down while the other is drawn up; but they would be difficult to work, and the complicated operation would not save much time.

I am of opinion that if the scrapers were of soft timber, or even if made in the form of brushes, they would be sufficient to remove any small shells or weeds that were not allowed by neglect to grow too large, at the same time that the paint or other preservative coating on the plates would not be injured by their application. I have tried the scraper described, on a vessel in the tide-way of the Mersey, and found it to operate most effectually. Various means have frequently been employed to scrub the bottoms of wooden vessels while they were afloat; but the several plans were generally attended with great labour to the men, from the want of proper tackling; and they are now, in some measure, rendered useless by the introduction of copper sheathing. I trust, however, that all such clumsy contrivances may soon be rendered unnecessary, and that no such objection as liability to fouling may remain as a handle for those who oppose the employment of iron as a material for ship-building.

STIFFNESS OF HULL.

It is usually asserted that the stiffness of the hull is detrimental to fast sailing, and, to support this opinion, the fact is often quoted that smugglers and others when chased by an enemy and losing ground, as a last resource, cut the beams of their vessels, and otherwise loosen the frame-work, by which the speed is said to be increased. But I cannot help doubting much that I have heard on this point, although it is not easy, or perhaps safe, to deny the accuracy of an opinion that has been generally entertained by nautical men. As it is to be supposed that the form originally given to the vessel is suited to produce speed, it is scarcely reasonable to assume that any accidental and trifling alteration, thus caused by the waves, can be an improvement. May not the advantage gained by loosening the upper-works of a vessel arise from the play thereby given to the masts, reducing the jerking and straining which tight rigging is known to cause, and admitting a steadier and more even pressure of the canvass? The seamen in the "Baltimore clippers" never use shrouds, until the force of the wind obliges them to set them up, lest they should carry away a mast; and our yachting men well understand the advantage gained, in a closely contested race, by slacking the weather rigging.

LEEWARDNESS AND STEERING.

Two other objections to iron vessels are frequently made, which may be briefly and satisfactorily disposed of. An opinion prevails that they drift to leeward; and that they do not steer well. These are disadvantages solely dependent on form, there being nothing peculiar either in wood or iron that can affect the question. Hitherto the generality of iron vessels have been made for a light draft of water, and being consequently flat floored and very full, without adequate and counteracting depth of keel, they have not sufficient hold of the water, and steer imperfectly. The remedy for both defects is so obvious and simple that it is unnecessary to discuss it.

IMPROVEMENTS RETARDED BY PREJUDICE.

I cannot for a moment suppose that those who are capable of forming an opinion for themselves, and have had opportunities of judging from experience of the comparative merits of wood and iron, and are willing to be convinced—I cannot suppose that in their minds any doubt remains, or that they can be deterred from pursuing the subject, by the groundless objections that have been raised against vessels built of iron. But I would rather address my remarks to those who have not had such opportunities, or have avoided the enquiry. To be convinced of the impolicy of overlooking improvements merely because they do not accord with their pre-conceived notions, it is but necessary to look back for a moment to the numerous instances in which the most important and even palpable improvements have been checked and slighted by ignorance and prejudice, and we shall perceive that the public is not an infallible judge of matters that strike at the root of old and inveterate impressions.

One very remarkable instance of this kind is within the memory of many, and is connected with our present subject. When chain cables were first introduced, the greatest outcry was raised against them by those of the old school. A ship provided with these only could not then be insured. At a subsequent period, it became the custom for ships to have one cable of hemp and another of iron. But where now is the well-found ship that has a hempen cable? At present, iudeed, it is a question whether the underwriters would not be inclined to dispute the payment of a loss by a vessel in which iron cables were not used. Iron rigging has, also, for some time, been struggling against the opposition of seamen, who view innovations with an extraordinary degree of abhorrence; but, since the introduction of wire rope, I have no doubt but all standing rigging will shortly be made of that material, its superior strength and power of retaining its tension, as well as its requiring, from its small size, much less displacement, rendering it preferable to hemp.* Iron running rigging in chains, for halliards, topsail sheets, &c., is also extensively used, and is found to be superior to hempen cordage for these and other purposes, being much more slender than ropes of the same strength, so that it will pass through smaller sheaves or blocks, while, at the same time, it stands better to its work from its not being subject, in atmospheric changes, to any perceptible expansion or contraction.

The rise and progress of steam navigation are frequently referred to as furnishing instances of the effect of public prejudice operating against great national improvements; and it is rather amusing to read the reports of the evidence given in the House of Commons about the year 1822, when steam packets were first established at Holyhead. The most sanguine man could not, from these documents, at all picture to himself the effect that twenty years would produce in that department of shipping; yet they contain the evidence of all the most experienced men of the day.

---

* On this point I quote the following paragraph from a recent number of the *Mechanics' Magazine*:—"IRON WIRE RIGGING.—A vessel, the '*Marshall of Grimsby*,' is now in the river off Hermitage Stairs, the rigging of which is all of wire, (Smith's patent wire rope.) It has been nearly six years thus fitted up, and the rigging is stated to be nearly as good as new.—Taking a 3-inch wire rope, and comparing its size, weight, and cost with hempen rope, or chain, of equal strength, the results are found to be—A 3-inch patent wire rope, weight $6\frac{3}{4}$ lbs. per fathom; hemp rope, of 8 inches, $14\frac{1}{4}$ lbs.; and 13-16ths chain, 36 lbs.—equal to a strain of 16 tons. The comparative cost being—3-inch wire rope, 3s. 8d. per fathom; 8-inch hempen rope, 6s. 3d.; and 13-16ths chain, 7s. 6d. per fathom.

IRON VESSELS, AS APPLIED TO STEAM NAVIGATION.

If the arguments adduced in favour of iron for ship-building be correct, we may date from the present time an era in the history of navigation most important to the world at large, and more especially to this country. To steam navigation, its importance can scarcely be calculated. The machinery and fuel occupy so large a portion of the vessel that very little room is left for stowage. The weight and tremour of the engines shake and strain wooden vessels, and the machinery itself, until the repairs of both swallow up most of the profit that would otherwise be realized; and it is no trifling consideration that, in using iron for the hull, a finer form can be given, and, consequently, a much higher rate of speed than can be attained in wooden vessels, while the lightness of iron vessels renders less necessary the full bow that is now deemed an essential to the sea-worthiness of a timber-built ship. It may be added, that some of those nuisances which (though considered comparatively trifling by seamen,) are exceedingly unpleasant to sensitive voyagers, are nearly, if not entirely, unknown in iron vessels. The nauseous smell arising from bilge-water, vermin, worms, and the eating away of the timber are instances of this kind, and their absence is very advantageous, especially in tropical climates.

It has long been felt to be desirable, in steam navigation, that a greater degree of adaptation should exist between the vessel and the machinery. In wooden steamers, two interests have always to be reconciled. The ship-builder employs all those means which he conceives to be best calculated to strengthen his vessel, and too often clings to principles of construction, which, when applied to steamers, are ill suited to the attainment of high speed. The engineer, generally very regardless of the requirements of the vessel, and having little sympathy with the ship-builder, thinks only of seeing his engines well accomodated, and working smoothly. If the vessel, when completed, goes well, both are exalted; if the reverse, neither is to blame. Unfortunately, a fair excuse may be made for this want of unity of design in the two departments. A wooden vessel, when she takes the ground, and, frequently, when she encounters heavy weather, becomes so much altered in shape that, were the engines so attached as to form "part and parcel" of the whole, they would be in danger of being broken down, or otherwise injured. Few wooden steamers, I may add, can be placed in the graving-dock without first slackening the holding-down bolts of the engines, to allow the vessel to spread out, lest the framing should be broken. The case is wholly different with iron vessels—in which, if properly built, there is no fear of twisting or settling. The engineer may, with perfect safety, firmly unite the vessel and the engine. Having to design both, and being responsible for the success of the whole, he will not allow one part to suffer for the benefit of the other, but take an interest in so combining both that they may mutually support each other. The various resources with which his profession furnishes him, enable him to carry out one uniform and consistent plan. The lines, the trim, the draft of water,† the strength, the power—all may be adapted to each other, and a degree of excellence be attained of which we can as yet form no adequate idea. Then, indeed, we may expect to see a complete steamer.

Let us again direct our unbiassed attention to wooden ships, and, viewing the subject dispassionately while we examine them, we shall be struck with surprise at the means employed in their construction, and the extraordinary complication of the frames, especially in large ships. It has been the work of ages to determine upon the several systems adopted, yet it is well known that one and all in them are still very far from being efficient; and nothing proves more clearly the difficulty that surrounds the question, than the fact, that, to this day, with all the time, talent, and experience that have been brought to bear upon it, it is by no means decided which peculiar principle of building is the most advantageous, and it is within only the last few years that a total revolution has been made in the science,—affording internal evidence that timber is *deficient* in many of the qualities requisite for ship-building. The force of custom is indeed strong; and had we throughout adopted iron alone as a material, it would have been naturally thought as impracticable to employ timber for such a purpose, as it now is to use iron plates.

As a proof of how much this is required, I may instance the case of a timber steamer which drew about two feet six inches more water, when her engines were put in, than was contemplated,—an error of probably 200 tons in the total weight. Very few, indeed, approach nearer than 6 or 12 inches to the line originally intended. This may, in some measure, be attributed to the uncertainty as to the specific gravity of the timber to be used. With iron the draft may be attained with greater certainty. I have, myself, calculated the draft of water of a steamer of 470 tons measurement before her keel was laid, and when she was completed with engines, and a given weight of cargo, no perceptible variation from the depth named could be discovered.

I have before alluded to the powerful influence possessed by underwriters in causing the public to encourage or reject improvements; but in the same proportion as their unfavourable opinion has impeded, and still may for a time impede, the building of iron ships, so will their approval and willingness to reduce the premiums of insurance on such vessels cause them to come more rapidly into general use. I sincerely hope that this desirable change in the sentiments of underwriters is not far distant, and that they will render unnecessary what has long been contemplated by the iron masters, and other men interested in iron vessels, —namely, the establishment of a Marine Insurance Company that will insure iron vessels at moderate premiums.

Mr. Holmes, in his report on the steam communication with India, has expressed himself much more decidedly than I have done, and seems to expect greater advantages from iron than I have attempted to claim for it. He thus writes :—

"Iron ships are not so liable to be wrecked: they might, if properly built, be run on shore, or on the coral reefs, or stranded, without sustaining any material injury: they can be rendered truly safety ships, so that their sinking, or any separation or opening of their parts (as is the case in a wreck, or as in the late calamitous loss of the "Forfarshire" steamer) could never occur. This advantage alone should be sufficient to claim a preference, when it is considered that at least 500 British merchant ships are annually lost at a great sacrifice of life and property, besides those belonging to the Navy.

"An iron ship is not liable to injury from hot climates, lightning, or fire. The whole is kept cool, free from vermin, and in a healthy condition for the crew. It is not liable to leakage, so that no damage can arise from bilge water to the cargo, or to the crew from unhealthy odour.—How important to the corn trade!

"In wood ships, a considerable elasticity exists throughout the structure. This elasticity and the motion of the machinery work one against the other with an alternate action that has a tendency to destroy each other, so that the best arranged and strongest machinery is never to be depended on, as an extraordinary twist or strain of the vessel may commit damage in a single moment which the regular work of years could not effect, besides putting all to a stand at a time when the utmost efforts are wanted, as the most likely time for an accident of this kind to happen would be in unfavourable weather, and with the wind a-head. The iron ship, on the contray, is as one solid block or mass, not admitting of the slightest twist, or hog, or change in its position, or truth in its parts one to another. The steam machinery may be fixed immoveably in it without any chance of its various parts working loose, or becoming twisted. It can work with all the advantages as if fixed on a *terra firma* foundation, and the whole ship might be placed on a truck and drawn over the stones without injury; while a wood ship, if put to such a test, would be shattered to pieces.

"In working steam vessels against a gale, the machinery can only be exerted with great caution, without risking the destruction of the ship. Such is the effect of the alternate action of the ship and machinery against each other that if the power of the machinery was not to be reduced in a gale, (instead of being exerted to its utmost when most required,) the leakage would sink her; and, in many cases, no ordinary-built ship, if even secured against leakage, could hold together. But an iron ship would sustain no injury in working its full power in any gale, and would be too buoyant to work under the waves."

IRON THE MEANS OF EXTENDING STEAM NAVIGATION.

In contemplating improvements in steam vessels, we are not to confine ourselves to a few advantages that may be given to lines already established; but, embracing a wider range, suppose them constructed of materials that will remain firm and wholesome for many years in any climate, and that may without much danger track through seas beset with hidden rocks and treacherous coral reefs.

If not injured by some casualty, a strong iron vessel will not require to be put into dock for several years, provided opportunities are occasionally afforded for leaving her dry for a few hours, to clean and coat her bottom. The engines being reduced in size and weight; the speed and stowage much increased; the consumption of fuel lessened to one-third of the present average,—what part of the globe may not then have its steamers, what sources of employment does not the prospect hold forth to this country, and what immeasurable advantages may not thereby arise to our widely extended colonies! When we consider the results of such improvements, we are lost in conjecture of the amazing extent to which our resources may be applied. We have steam vessels now plying regularly to North and South America: we have them in the Pacific, in the Mediterranean, in the Red Sea, and in the Indian Ocean, as far as Calcutta, and even China; and these are only the main navigable highways, from which numerous small lines may be expected to branch, filling up every intermediate space, affording a continuous and profitable intercourse with countries the most remote, and thus diffusing amongst tribes the most barbarous or benighted, the blessings of civilized life.

Our steamers may probably ere long do more to keep open and extend our trade with China, by plying along the shores and up the rivers of that country, than can ever be accomplished by the sound of our guns. South Australia—the antipodes to our own country, and in itself a continent—also presents itself with even higher claims to our consideration as the mother country; and it is not too much to assume that there the general adoption of the English language, and English iron-built steamers, will concurrently be the promoters of commerce and civilization. All these redeeming advantages, and many more, may be achieved by British enterprise. But, knowing, as I do, from long experience, the precarious and deteriorated condition of timber-built steam vessels, when long exposed to such service, even improved as they now are—knowing the constant necessity for repairs, without the means of effecting them, and the comparative want of room for stowage, I feel satisfied that it is left to the employment of iron, and all the improvements that will follow in its train, to carry out those desirable objects.

IRON VESSELS IN TROPICAL CLIMATES.

It has been asserted that iron vessels become too warm in tropical climates, and that they are consequently unhealthy. How this opinion arose I cannot divine. Every report I have read speaks strongly to the contrary. I believe that iron steamers in the East have been highly approved of; and the same reasoning that applies to merchant vessels, on this point, is equally applicable to ships-of-war—a fact of great national importance. The iron war steamers now employed on the coasts of China are said to be exceedingly agreeable to the crews; and Captain Trotter, who headed the late unfortunate expedition to Africa, far from attributing the fatality amongst the crews under his command to the material of which his vessels were built, expressly states, that, if the enterprise is to be pursued, he should give the preference to iron-built ships. The probability indeed is that, had they been of wood, the mortality would have been greater.

IRON VESSELS, A NATIONAL QUESTION.

There could not be a better period than the present, when the the foundation of every commercial system seems to be shaken, and men are looking around in dismay at the depression that weighs down every branch of trade—there could not be a better period for considering with deep attention a question such as this, and endeavouring to draw from it some assistance towards alleviating the general distress. As a *national subject*, it will be found to possess more claims to our attention than are at first apparent.

To do full justice to this enquiry, it would be necessary to consider all the complicated questions relating to the timber duties, and the effect which the substitution of iron for wood, in ship-building, would have upon our colonies; but I confess myself unequal to this task, involving, as it does, many nice distinctions, and much tedious research. But, difficult as it may be to make correct calculations on this subject, and to show by figures the precise relative values of all the points connected with it, no one will deny that it is of immense importance to this country that her ships should be built in her own ports, with materials of her own production, and procured by the labour of her own population. And it must add greatly to the importance of this consideration, that the most suitable materials are abundantly found in the deep recesses of the earth, leaving the surface undisturbed, to be employed more exclusively in the growth of corn instead of woods and forests. I would not be understood, by this argument, to admire the taste of our friends in America, who, after emerging from their dense and boundless forests, run into the opposite extreme, and see so much to be desired in their equally extensive prairies, without even a shrub or a twig to relieve the endless monotony of the scene. On the contrary, I feel all an Englishman's pride at the sight of a noble oak, which forms at once a shelter and an ornament to the land on which it grows. But how often are these swept away just as they attain their full growth and beauty, under the irresistible temptation presented by the great demand and consequent high price of the timber!

At the last meeting of the British Association, at Plymouth, Mr. Chatfield, of her Majesty's Dock Yard there, read a paper on ship-building, and gave some interesting details respecting the "Hindostan," seventy-eight-gun ship, then on the point of being launched. He stated that,—

"To construct a ship of the tonnage of the "Hindostan" is a work of great labour, time, and expense. The value of labour alone is about £12,000; and it would take 154 men twelve months to build her. The value of materials is about £60,000, making the cost of the ship £72,000 sterling. It would require 4200 loads of timber to build a ship of that description, and occupy 70 acres of ground 80 years, for its growth.* The average durability of ships-of-war, employed in active service, has been calculated to be about thirteen years, when built of British oak, which happens to be precisely the period the "Hindostan" has been building, for she was commenced in August, 1828, and will be launched in August, 1841."

I select this case as being of recent date, and coming from an unquestionable source. It proves, first—the great space of ground and the long time required to grow the timber of which they are built; and, secondly—their short average durability. Taking the facts stated as our data, but deducting one-fourth from the quantity of timber required for the "Hindostan," as an allowance for the lighter scantling of one of our ordinary merchant ships, it would require the constant occupation of about 400,000 acres of land on which to grow the timber for the ships annually built in this country and in our colonies. So much of the timber thus employed as is grown in

---

* Equal to 5600 acres for one year.

this country occupies a proportionate amount of land that would otherwise be appropriated to agricultural purposes; so much of it as is not grown here, takes our capital, to a certain extent, out of the country; but, worse than all, our people are deprived of profitable and extensive employment by the inducement to build ships in those countries where the material is most abundant. By the evidence of Mr. Meek, who was lately commissioned by government to enquire into the state of trade on the continent, in relation to this country, we learn that it is impossible for our ships to compete in freight with those built by several other nations, owing principally to the comparatively low rate at which they can procure the materials for building. Notwithstanding the great sacrifice now being made in the revenue of this country, by the reduction of duty on foreign timber, and the consequent decrease that will probably be made in the price, the foreigner will still have an advantage over us in having the material not only considerably cheaper, but at his own door, while our supply will necessarily be precarious.

STATE OF SHIP-BUILDING IN ENGLAND.

And here one cannot reflect on the state of English ship-building without perceiving that some great alteration is necessary in order to place it in its proper position. The price of new ships is too much reduced to allow room for profit, when labour and timber are so expensive as in this country; and I believe it is generally acknowledged that, since the war, our ship-builders have mainly depended for subsistence on repairs alone. But is it right, I would ask, that a large and respectable class of men whose profession should, from its importance, enable them to rank as high in wealth and station as they do in intellect—whose occupation requires the attainment of great scientific skill, should have their talents thus degraded—their energies thus cramped? Is it right that, in a land whose merchants are princes—in a land which claims the boundless ocean for its empire, the men who have spent their lives in so important and national an avocation should have no fair remuneration for their services? And where can we see any prospect of improvement while timber alone is employed, and while our population (already so dense) and the circumscribed limits of our soil for ever deprive us of the power to grow timber in sufficient quantity for our own ships?

But let iron become the material with which our ships are henceforth to be built, and the whole question assumes a widely different and a highly cheering aspect. Without being in any degree dependent on foreign countries, we should find an inexhaustible supply of more suitable and less perishable material for the whole of our national and mercantile marine in our own country; from this source our iron-masters would have a fresh and a steady demand for their iron; and an increased demand for labour, both at the mines and in our building yards, would be the immediate and invaluable result.

If I have not over-rated the superiority of iron in what I have advanced, it is clear that few foreign timber-built vessels could compete with our iron ships; and if other countries are driven to the use of iron, there can be no doubt where the advantage will lie. All nations yield the palm to England in the production and working of iron, and it will be long before we can be deprived of our superiority in this respect.

Our navy would be benefitted to an equal, or perhaps a greater extent than our merchant service, by adopting iron, and this consequently renders the subject one of still greater national interest. I have shown that greater strength and higher durability may be attained by iron than by wood, and *that,* too, in a rapidly increasing ratio, as ships increase in size; and I have also suggested means by which unlimited strength may be given in a simple and obvious manner. On this point, Mr. Creuze's argument is at once strong and explicit. In his remarks on iron ship-building in the *Nautical Journal,* he states—"The answer to the second question, as to the limit in the size of the vessel to which the substitution of iron for timber may be carried, appears also to be involved in the foregoing considerations. For, if greater strength may be obtained with equal weight of material, or equal strength with less weight of material, there can be no limit short of that limitation which may equally apply to timber. And, by an application of the foregoing reasoning, to the question at issue, it appears that a first-rate *may be more strongly built of iron than of timber*, with the same light displacement; and equally strongly built, but capable of carrying a greater quantity of water, provisions, and stores, with the same load displacement; or equally strongly built, and capable of carrying a quantity of water, provisions, and stores, with a less load displacement. This may appear to be a bold and startling result of our investigation; but if our original assumption be correct, it is, nevertheless, within the bounds of truth. Nay, it is even an *under estimate* of the limit to the substitution of iron for wood in the construction of ships. For, the limit to the possibility of constructing a fabric of any conceivable dimensions, is necessarily dependent upon the ratio of the strength of the material used to its weight. And, as this is greater in iron than in the timber of the timber-built ship, the limit of dimensions for the iron-built ship is more extended than the limit of the dimensions of the timber-built ship."

These arguments of Mr. Creuze will ere long be fully verified. Who, that has seen the noble steamer at Bristol, or the delicate and elegant yacht of our celebrated townsman Dr. Brandreth, can avoid the conviction, that iron may be worked to any size, or to any form, with all the regularity of the most beautiful timber-work; combined, however, with a degree of strength which in no timber-built ship can ever be attained?

EFFECT OF SHOT ON IRON VESSELS.

It would be taking a contracted view of the question to suppose that we have at all approached the extent to which the strength of iron plates and framing may be carried; or that we may not, in the progress of improvement, even construct a body that will almost defy the force of a cannon ball. It would, indeed, require a very great mass of iron to accomplish this; greater, perhaps, than would in practice be found desirable; but I fearlessly assert, that an iron ship of war, may, without difficulty, be rendered much more effectually proof against shot than a timber one—probably to the extent of resisting all but point-blank shots. If this should be proved to be the fact, we shall prevent all the injury and loss arising from oblique shots, which, under some circumstances, are exceedingly destructive in wooden vessels. The loss of life from splinters, would, moreover, be entirely avoided. I need not attempt to picture the fatality, accompanied by wounds of the worst description, arising from this source. The loss of life occasioned by the shots themselves is estimated at from one-third to one-sixth only of the whole—all beyond that proportion being the result of splinters.

We will now consider the comparative effect on the vessels themselves. Judging from my own observation of the injuries arising from severe collision with hard bodies, I feel quite satisfied that, in almost all cases, an iron vessel is more capable of resisting a concussion, even in the part struck, than a timber-built vessel; but, in addition to this advantage, the injury in the former is always confined to that spot, while in the latter, the seams and frame-work are shaken throughout. I therefore conclude that, though an iron ship of war might not be proof against a shot, the effect produced would be simply a ragged hole the size of the ball, and the indentation of the part close round it. Beyond this, it would cause no injury whatever. Shots striking the hull in a slightly oblique direction, would probably only indent the plates, and fly off; and all others (more obliquely directed) would be comparatively harmless. Any hole made in the vessel could be readily approached from the inside, and plugged up, if necessary. Proper tools, plates and bolts, should be provided; and a few mechanics, with the assistance of the crew, could very soon repair any damage that might thus arise.

I have somewhere seen a small instrument, which, although simple in itself, might be rendered applicable to the stopping of shot holes in an iron ship, especially as the part injured can always be seen from the inside. It is made like an umbrella in miniature with a long iron handle. The arms are also of iron, and are covered with very strong canvass. When it is closed, the point is thrust into the hole, till the part which is made to expand projects entirely through. The handle is then drawn back, so that the arms and canvass spread out till nearly flat, and cover the hole. The pressure of the water from without will keep this nearly tight, and the handle may, at the same time be drawn firmly in, and secured to the frame of the ship.

The facility for stopping shot holes in a wooden ship is by no means so great. In the first place, many more shots would take effect—frequently shattering important knees, beams and frames. The external hole must be approached from the outside; but the timber having partially closed the aperture, it is difficult to insert a plug in an efficient manner. The injury, moreover, is not confined to one spot. After a severe action, the greatest exertion and ingenuity are frequently required to keep a ship afloat, owing to her general great leakage ; and many vessels so shattered, though immediate danger may be averted, have to be sent home for repairs.

Teak, which is the strongest and most durable timber for ships-of-war, is the most liable to splinter, and is on this account much dreaded by seamen. One shot and the splinters occasioned by it have been known to kill and wound no fewer than nineteen men.

The action on a vessel, arising from the recoil of her guns, is much in favour of iron, if we may judge from the effect produced on the "Nemesis" and the "Phlegethon," each carrying two thirty-two pound pivot guns, and also fitted with twenty-four pounders. They have sustained no injury from the recoil of these heavy pieces of artillery, while it is said that the timber-built steamers employed in the same service were shaken, and became leaky.

OUR STEAM NAVY.

The steam department of our navy now requires notice, and I do not shrink from advancing some observations upon it, although I may thereby expose myself to the charge of being somewhat presumptuous. It is well known that, in this department, the operations of the government for several years lagged far behind those that have resulted from private enterprise. Until within a recent period, they seemed to have viewed steam as an innovation of questionable utility upon old-established practice and national feeling, as many yet regard the introduction of iron in ship-building ; and they appeared to pause, with an almost unaccountable apathy, till the great national experiment had been tried at the cost of individuals. I need not ask whether such a course was to have been expected from the government of a country where all is dependent on her ships. They would surely have done well, had they called together all the united talent of the nation, and stimulated the energies of her engineers. To the want of such steps may, in a great measure, be attributed the circumstance that steam property has been the ruin of thousands. Owners of steam vessels, narrowed in their resources, were obliged to trust for improvements to the random experiments of unscientific men, who have had no standard to guide them ; and the burden, which would scarcely have been felt, had the well-combined exertions of a government been directed to the subject, has been allowed to fall upon her people. To say that two per cent. has been realized, on an average, on the capital expended in steam property, is perhaps assuming too high an estimate; and thus one of the greatest triumphs of modern genius—one of the most surprising and gratifying results of scientific skill that the world ever witnessed, has been an injury and a loss to those who have achieved it. I thus strongly allude to this fact as a proof of the bad effects, in a national point of view, of a want of sympathy in the government with the people in promoting improvements. I sincerely hope that a different course will be pursued with respect to the subject I have now undertaken to advocate; and I respectfully, but earnestly, call upon those who direct departments embracing this great question, to come forward and give it that fair and ample investigation which, I feel convinced, would enlist them in its support. It is worthy of every consideration from this country, especially; and should not be left to the mercy of crude conceptions and vulgar prejudices.

A fleet of very large steam ships-of-war is now in course of erection, and I do not hesitate to predict that there is not a man now employed in their construction but will ere long regret that they were not all built of iron. Sufficient evidence on all essential points can already be obtained to guide us in forming a correct opinion, and those who overlook such testimony will have cause for self-reproach. A decided improvement would be effected by the adoption of iron in every description of our shipping; but I know of none in which the result would be more beneficial than to our steam navy. These ships require the greatest possible amount of stowage room, to enable them to carry an adequate quantity of coals and stores for long voyages; they are sent, at great expense, to distant stations, where it is desirable they should remain as long as possible without coming home for refitting or repairs. It is necessary they should be cool and healthy for their crowded crews, in hot climates; and it is of great importance that they should be calculated to withstand the recoil of their own guns, and the shot from an enemy. Those who feel satisfied with the arguments I have adduced will readily admit that all these results may be obtained by the use of iron.

So long as the present paddle wheels are used, no steamer can go into close action. The engines may therefore, I believe, be sufficiently protected from shot, which are generally distant; but, should a sufficient substitute be found for paddle wheels, and the vessels be enabled to go to *close quarters*, means may be employed *effectually* to protect the engines from injury.

There remains, then, but one obstacle to the general introduction of iron in ship-building, but that is certainly one which in this country it will be always difficult to overcome,—namely *national prejudice*. We cannot easily bring our minds to relinquish the "wooden walls of old England," which have done the state such good service, and which need but to be named to command respect in every quarter of the globe. Yet, why should we despair of finding popular prejudice give way to popular conviction in favour of our "iron walls," as the means of perpetuating the maritime glory of our country! However much this national feeling is to be respected, we cannot long close our eyes to the evidence of facts. It is well known that many of our most magnificient specimens of naval architecture are impregnated with the seeds of decay in every beam, and entail upon the nation enormous costs in continual repairs, as well as the frequent necessity of constructing new ships. But, whatever change may be made in the material of which the ship is built, Englishmen may feel assured that the *real* hearts of oak are still unchanged, and that they will ever be found ready to unfurl her sails, to direct her course, and to apply the match when the enemy is at hand. Recent events too clearly shew that no dry rot can ever find entrance there!

By reference to a paper, before mentioned, written by Mr. Creuze, we find it stated that *fifteen* years is considered to be the average duration of ships of war. Mr. Chatfield, indeed, makes it only *thirteen* years, while in actual service; and although many ships of a much greater age are now in existence, these are sometimes virtually *new* ships,—since, after a few heavy repairs, very little of the old timber remains, and the same parts of a ship are frequently two or three times renewed. There are, however, some ships of a great age, having much of their original timber left; but these are only exceptions, and may be set off against a much greater number of pine-built ships which have only lasted five or six years. Numerous instances may be adduced of the loss sustained by the nation from the ravages of the dry rot, without taking into account the extra number of vessels required to fill the places of those under repair. I may instance one case out of many. A splendid 120-gun ship, which cost the country £120,000, required twice to undergo the most extensive repairs, without having been once in service. There is now at the quarantine ground, in the Mersey, employed as a hulk, a beautiful frigate called the Arethusa, which, from the early development of the dry rot, was condemned before she ever put to sea! Iron vessels, on the contrary, built of a corresponding strength, would doubtless last for *fifty* years, or longer, with very light repairs, and these at very remote periods; and, the durability of an iron vessel being once ascertained, would be a fair standard for all others of the same thickness.

To pursue this portion of the subject, would only be to enumerate the advantages I have before described; but great as those are to our ordinary merchant ships, the value of iron would be still more apparent if employed for the navy, and for all large vessels; and it is much to be desired that government would turn their attention to the subject, and order the building of some large iron ships, so that the qualities of the material might be fairly tested. I feel convinced that were this done, the Admiralty, following the example of private individuals, whom experience has brought to a slow but certain conviction of the superiority of iron, would never return to wood in the construction of our ships of war.

GENERAL REMARKS.

I am aware that I have not enumerated all the points that might be urged in favour of iron, or all the objections that could be arrayed against timber, as materials for ship-building. I am sensible too, that there are many abler heads which might have done more justice to the subject; but, at the same time, I conscientiously feel that I have not intentionally, even indirectly, exalted the one unduly at the expense of the other.

I cannot doubt but that many will receive my remarks with suspicion, as emanating from one whose judgment may be considered to be warped by prejudice or self-interest. I have but to implore those who may be impressed with such a feeling to follow my example —to enquire attentively for themselves—assured, as I am, that such an enquiry will lead them to a daily increasing confidence.

Can there be, I may ask, a stronger argument in favour of iron than the fact, that those who *have* thus acquainted themselves with its capabilities, and, either from necessity or conviction, have adopted it, have, almost without exception, been loud in its praise, and have shown their sincerity by not returning to the use of wood?

Imperfect as, I am aware, my observations are, I trust I have said enough to direct those who are resolved to investigate the subject, to the great points of difference between wood and iron for ship-building, and to assist them in arriving at accurate conclusions. Let me also hope that I have succeeded in impressing upon the minds of those whose interests are concerned, but who have hitherto evinced an extraordinary indifference to the advantages of iron, the prudence of instituting *some* enquiry, at least, rather than, by their apathy, continuing to interrupt the slow but sure current of improvement, of which they might take immediate advantage, and be amongst the first to share the benefits likely to result from this improvement.

I trust I have said enough, too, to show that it is the imperative duty of the directors of the large steam-boat companies of this country to enter into the enquiry with that energy and firmness which will enable them to arrive at the truth. I need not remind them of the serious responsibility they will incur, and how much right the shareholders will have to complain, if they do not. The practical results already attained by the introduction of iron, leaves them no excuse even for delay; and the position of several of the steam packet companies should operate as a stimulus to immediate investigation. These companies have expended large sums of money in the construction of new vessels, many of which are built in the best manner that modern art can suggest. No expense is spared. Wherever it is thought that copper or iron fastenings can be advantageously placed, they are adopted. The greatest care is taken in the selection of timber, and every means applied to render them not only strong and serviceable, but durable; and it is, no doubt, contemplated by the projectors that many of their mighty vessels, now afloat, will stand in the first rank for at least ten years, and be good serviceable ships for ten years longer. But what will be the consequence if, in two, three, or four years, it should be universally acknowledged that iron vessels possess all the advantages

I have claimed for them? What will then become of those steamers on which £60,000 or £80,000 have been expended? It would not be easy to remove them to other stations—for, who would have them? Even at half their original cost, they would be unsaleable; for, with an extra £10,000, or with the sum of £40,000 or £50,000, an iron vessel could be built, having equal advantages for carrying, considerably more speed, and capable of being worked at one-third less expense!

This fact is of the greatest importance in reference to the performance of all long voyages by steam; and it may be fairly asked, can any company, established for this purpose, pursue, at present, a prosperous course, unsupported by government funds? Very large sums are now annually paid to several companies for carrying mails and despatches which would (in each ship) scarcely displace a ton of cargo, or a single passenger; and yet what are their prospects? It remains to be proved whether they are a good property, and whether they are likely to become profitable investments. But suppose the same work was to be performed by vessels of two-thirds the first cost, worked at two-thirds the annual expense, requiring less than half the repairs, having the same accommodation, and greater speed—all of which would be attained were iron vessels employed—what would then be the position of such companies? The receipts, which, in the one case, would scarcely pay a low rate of interest on the capital, would in the other yield a large profit.

I may here refer to another passage in the Report of Mr. Holmes, which bears on the subject under discussion. He says:—

"Iron ships are especially calculated for that vast revolution in the navigation of the world which is at hand, by steam vessels superseding every other navigation or mode of sailing. Without iron ships this could never be, from the difficulty of building very large ships of wood to draw little water, with a due regard to those principles of naval architecture which are indispensible in their structure. The consequence would be, that sailing vessels never would be superseded by steamers for the conveyance of general freights, but for iron ships. Hitherto, wood ships of an ordinary build have been as large as may have been deemed manageable under sails. There has been the limit. But where will be the limits of steam power? The only limits to the size of iron vessels are the draft of water and extent of traffic. As far as relates to the strength, flotition, and control, there is no limit.

"Such, indeed, are the advantages of iron ships over those built of wood, that they only require to be known and understood to be universally adopted, to the exclusion of all others. No wood ships can compete with iron ships in profit or accommodation; and, wherever wood ships are now employed, iron ships may supersede them, and defy competition, even with old established trades. This must be the case even with the "Great Western," "British Queen," and "President" ships; they will hardly be established on their stations before they must be removed. I fearlessly predict this to be the case, and that, after a few years, no large wood ship will be built in England, either for mercantile or naval purposes. I make these statements after a mature and deliberate consideration. I feel that I may be saying more than can be at once admitted by your committee; but whether they may establish confidence or distrust in my opinion, I feel that I should be wanting in my duty, both to you and the committee in Bengal,* were I to express less than I have done, having the fullest conviction that, were you to build wood ships, you could not work them longer than you could replace them with iron; consequently great loss would be the result."

VALUE OF OLD VESSELS.

The conclusion of my paper is not an unfitting place in which to speak of the result, when (and the time must come) both *timber* and *iron* ships shall be pronounced no longer sea-worthy, and must necessarily be broken up. In comparing their respective value at such a crisis, the copper of the wooden ship should not be taken into account, inasmuch as the cost of that portion of her material is entirely saved in the first cost of any iron vessel.

The hull of a timber-built ship, when condemned, would probably not produce more than ten shillings per ton of her register tonnage; while the iron vessel—taken at ten cwt. of iron for every register ton, and valuing the iron at the rate of £4 per ton—would

* Extracts from Mr. Holmes's Report on Steam Navigation to India.

produce £2 per register ton, or four times as much as the wood ship.† This corresponds with the statement of the friend whom I have quoted, with respect to canal-boats in Staffordshire, and who says:—"When worn out, iron boats will sell for about £25 each, and those of wood, for only £4 or £5.

In conclusion, I must crave the considerate indulgence of my readers. I am aware that I have exposed myself to some criticism by venturing to discuss this subject; but I trust my observations will be received in the kindly feeling in which they are written, however strongly I have found it necessary to express myself on some points. It must be recollected that I have had no guide; that it has devolved upon me to lay the subject, for the first time, fairly before the public. I am open to conviction, if I have overrated its importance; or shall again be ready to take up my pen as fresh proofs may arise, or fresh information may be given me, that may appear available for the public advantage. I am willing to discuss, publicly or otherwise, all or any of the points to which I have called attention. I will reply to any questions that may be put to me, and endeavour to meet any objections that may be preferred to the opinions I have advocated. I sincerely thank those who have followed me through the several arguments I have employed; and may I be permitted to hope that I have advanced the subject something in their estimation. May the period not be distant when all will be found cordially to agree in the opinion, that in every stage, and under all circumstances, timber ships must yield to iron, in their rise, their progress, and their end.

Since writing the above, I have reason to believe that these estimates are both too high; but the iron would appear to still greater advantage, were I to make any alteration.

APPENDIX.

(No. 1.)

Manchester, March 21st, 1842.

My Dear Sir,—My time has been so fully occupied since I had the pleasure of seeing you, as to prevent me entering upon the subject of iron vessels. The enquiry is of such importance as to demand the most deliberate consideration; and, although my attention has been directed to the subject for the last twelve years, yet I can only, on this occasion, give you a few facts which I have derived from my experience in iron ship-building.

You are aware that, two years ago, I submitted to the British Association for the Advancement of Science a long series of experiments on the comparative values of iron and wood as a material for ship-building. These experiments I am still continuing, and I hope shortly to lay before the public convincing proofs of the strength, security, and durability of this invaluable material.

That iron is a stronger, safer, and more durable material for ship-building there cannot exist a doubt, and that for the following reasons:—

1st. Because the strength of iron plates is to English oak as 49.65 to 10.01, being in the ratio of nearly 5 to 1.

2nd. An iron-built ship is safer, from the circumstance that the material of which she is composed not only admits of great strength, but, the joints being all united by rivets, effect a degree of security which cannot be attained in a timber ship. On the other hand, a vessel composed of wood has all her joints open; and, instead of the whole being united *en masse*, the sheathing or planking presents a series of joints which, in order to make water-tight, must undergo the process of caulking; and thus, although consolidating the parts by hard wedging, has nevertheless a tendency to start the sheathing, and probably cause a greater strain on the joints. In the iron-built ship these counteracting forces are not in operation, as the joints are the same as the solid plates, minus only the loss of strength sustained by the perforations for the rivets.*

---

* I have found by experiment that the strength of the joints of plates taken through the line of rivets are as their sections.—*Vide Extracts from the Experiments.—Athenæum.*

Lastly. An iron vessel is more durable and more serviceable than a timber one, because (unlike wood) it is not subject to the dry rot, nor do the repairs (as far as my experience goes) exceed one-third of those to which a timber-built vessel is subject.

In the use of iron for the purposes of ship-building, we possess many advantages which cannot possibly be obtained by any other material. It receives any impression, will mould to any shape, and, from its flexure, ductility, and strength, it is admirably adapted for the construction of vessels, where lightness and strength form the whole or a part of their constituents.

With a material of this kind, there cannot exist a doubt as to its value, and the great security thus afforded must become apparent, not only to the present but future generations, in the construction of our naval and mercantile marine.

I am quite aware that iron has been objected to on the score of oxidation, but I am satisfied these objections are not well founded, as past experience proves that an iron vessel, when kept in constant use, is not only free from oxidation but presents no more appearances of corrosion than railway bars, which are well known to retain their permanency so long as the carriages continue to roll over them. This principle of non-oxidation is equally apparent in the durable nature of the iron ship. If we keep her in constant use, there is no appearance of deterioration;—lay her up for a few months, and the usual appearances of atmospheric action become visible, accompanied with a rapid corrosion of the joints exposed. In order to protect the vessel in such a position, it is only necessary either to cover the plates with a coating of my friend Mr. Mallet's anti-corrosive solution, or, what will otherwise answer the purpose, two or three coats of good oil paint. The phenomena of iron ships, as well as railway bars, not being subject to oxidation, is an interesting as well as an important fact. Whether it arises from a vibrating action, or from some other repulsive property amongst the particles of matter thus acted upon, is not for us to determine. It is enough for every practical purpose that such is the fact, and that we have nothing to fear from causes which are evidently in operation when the vessel is constantly at work.

Another property of an iron-built vessel is, retention of form, and the great strength it exhibits under circumstances which, to wooden vessels, must prove fatal. In iron vessels taking the ground, it is not only known that they will yield to pressure, but they will, in many instances, by their elastic force, restore themselves to their original shape and form. This, in a great degree, arises from the nature of the material, and the combination of the parts, by which the whole, and more particularly the sides, present a powerful resisting force to rupture; whether arising from strains in the middle, when a vessel happens to be supported only at two points—the stem and stern—or from a reversed position where the point of support is under the keel at midships, and both ends are suspended in the air. In trying situations of this kind, the iron ship exhibits extraordinary powers of resistance. In almost every instance of this kind (and several have occurred) she retains her form, and not unfrequently escapes uninjured, and that entirely from the combination which exists in every part of the mass.

It was only the other day that I witnessed an iron steamer, nearly 400 tons burthen, in this peculiar position, with her stern in the air, and her hull at midships grounded upon a steep bank, with the sides covered with boulder stones. This took place last week on the Thames; when hauling the vessel out of a temporary basin, she grounded on the extreme end of the bank, and was left till the tide receded, with forty feet of her stern entirely without support, and her bow buried in the opposite bank. Now this vessel remained in the above perilous position, till the return of the tide, when she floated, and immediately afterwards proceeded on her voyage to Margate.†

Iron vessels have also been objected to on account of the supposed danger arising from their being driven on shore upon sand banks, or a rocky coast. In the first instance, it is supposed they would be knocked to pieces; and in the second, that holes would be knocked through their bottoms. There cannot exist a doubt but this would be the case, particularly in the latter; but in the former, or in both, *they would never go to pieces*, and hence arises the greater security in iron ships. It is evident, in a similar position, that a wooden vessel would break to pieces, and the whole on board would become a prey to the winds and sea. This would, however, never take place in a ship built of iron, as, however much the hull might be fractured and put out of shape, she would still hold together, and thus become a refuge to the shipwrecked mariner.

It would be superfluous for me to enter into other advantages of iron-built vessels, such as their superior lightness, capacity for cargo, &c. These matters are in better hands, and all I can offer is the few observations I have made; more as an illustration of my own experience, than any desire on my part to advocate iron in preference to wood. I cannot, however, close these remarks without adverting to the compass, which has always been considered a barrier to the extension and advancement of iron vessels. Professor Airy, in his valuable and truly philosophical experiments, has, however, set that question at rest, and I trust few will now dispute that iron ships can be steered with as much certainty as any other description of vessels.

I have thrown these observations hastily together. I fear they will be found of a very vague description, and probably of little value in elucidation of the enquiry you have at present under consideration. Should they be found at all useful, I must beg you to put them into a better shape, and oblige,

Dear Sir,

Yours sincerely,

John Grantham, Esq., Liverpool. W. FAIRBAIRN.

† Mr. Fairbairn illustrated this subject by a sketch, shewing the position of the vessel; but which, as the description will be readily understood, I have omitted.

(No. 2.)

Royal Observatory, *Greenwich*, 1842, *April* 8.

SIR, —I inclose the Manuscript Paper respecting the "Iron Duke." I had not heard of the loss of this vessel; and, indeed, had never heard her name. I cannot, therefore, from my own knowledge, give the smallest information as to the correctness of your surmise on the cause of the disturbance of her compass. But I think it extremely probable that the cause which you have suggested, or something similar to it, is the true one. When I was engaged in trying the "Rainbow," I remarked that the deviation of the sternmost compass, depending on the position of the tiller, was *seven degrees*. This however was much reduced when the compass was corrrected, (the observation to which I allude being made before the correction, and with the ship's head in that position in which the directive force was smallest,) but I cannot state from experiment to what it was reduced.

But, in cases of this nature, it is of the highest importance to examine whether some alteration has not been made, or some new iron introduced, or whether the place of the compass has been disturbed since the correction was made. The carelessness or boldness of Captains of ships in these matters approaches to something like infatuation. An instance of this kind was mentioned to me by Mr. W. Simms, (the eminent instrument-maker, of 136, Fleet-street,) a few months ago. The compasses in an iron ship, for which Mr. Simms furnished the magnets, were corrected by a competent person, (I think by his brother,) at one of the upper Docks in the Thames. The Captain proceeded as far as Gravesend, where he anchored, and returned to London almost in a state of frenzy at the erroneous action of his compasses. The person who had corrected the compass went instantly to Gravesend, and there found to his astonishment that between the time of correcting the compass and the vessel's sailing (a very short interval) a *large iron stove* had been erected close to the binnacle. Whether the stove was removed, or whether the magnets were re-adjusted, leaving the stove in its place, I do not remember: but in one of these ways it was arranged to the satisfaction of the Captain, and the ship has gone well ever since. This vessel might have been lost on the shoals of the Thames in her first voyage, and not a word would have been heard about the *iron stove*. I cannot give the name of the ship, or any other particulars, but I will refer you for them to Mr. Simms, who is by far the most proper person to give the whole history. As the story is so instructive that you may probably wish to hear it from Mr. Simms, I write to him by this post, stating that you will write to him for particulars, and requesting him to supply them to you.

With respect to the probable dulness of a compass needle in an iron ship, I can say, without hesitation, that there is no reason to believe the needle to be more dull in an iron ship than in a wooden-built ship.

---

‡ I have since heard from Mr. Simms, whose account of this circumstance is substantially the same as that given by Professor Airy.

With respect to the probability of any sensible change in the magnets, &c., experience is yet wanting. The ship in which the compass was first properly corrected is the "Ironside," (formerly the property of Mr. N. Cairns, of Liverpool,) which has made many voyages to Brazil, and the last time that I heard of her the compasses were still quite correct. Some of the next ships (I believe the next of all) were the iron ships running between Liverpool and Glasgow, whose compasses were right the last time that I heard of them, but upon these you can probably make inquiries if necessary.§ The next was the fast steamer the "Orwell," running between London and Ipswich, (a very intricate navigation;) I was on board her a few weeks ago, and she is still quite right. I have no doubt therefore that the correction may be trusted for many years; but it is very easy to re-adjust it at any time.

It has always appeared to me desirable that a proper department should be organised under the iron ship builders (as a body,) or under the Committee of Lloyds, or under some similar body, for the examination of the compasses of ships when they are ready for sea. This should be especially directed in the first instance to iron ships, but it ought ultimately to include wood ships, in which the errors are sometimes large.

I am, Sir,

Your obedient Servant,

John Grantham, Esq., Liverpool. G. B. AIRY.

(No. 3.)

Liverpool, March 14th, 1842.

SIR,—When I had the pleasure of hearing you deliver, at a meeting of the Polytechnic Society, an address on "Iron, as a Material for Ship-Building," I proposed to offer a few remarks, in the course of the discussion that ensued, favourable to your general argument, and based on my own experience on board of an iron vessel built by yourself and partners: the discussion, however, being prolonged to a late hour, I was induced to forego that gratification. But, hearing that you are about to publish the substance of your address, together with the opinions of several scientific men with whom you correspond, I feel that I should be doing injustice to my own feelings, as well as to your able advocacy of iron *versus* wood, did I not take this mode of communicating to you some particulars of a voyage in which the qualities of iron were tested in a manner that would give confidence to the most sceptical. I should premise, that although I have been a good deal at sea from time to time, and have sailed with some of the finest steamers of this port, I consider myself rather an amateur than a nautical man, and my own proper opinions must therefore be taken for as much only as they may be deemed

§ The compasses in these vessels have not been known to vary in the slightest degree.

to be worth; but, as the facts I shall adduce can be borne out by several practical seamen who were my shipmates on the occasion referred to, you are at liberty to avail yourself of them in your forthcoming publication, if you consider them in any degree calculated to forward the object you have in view.

Being anxious to view the beautiful scenery of the Shannon, I gladly accepted an invitation to accompany your new iron steamer, "Erin-go-bragh," from this port to Limerick last year.|| As this vessel was intended exclusively for the Lower Shannon, her draft of water was necessarily small, not exceeding $5\frac{1}{2}$ feet when in ballast-trim, and having but little depth of keel, she was not, of course, well suited for ocean navigation. She, however, performed the voyage most satisfactorily, though, for the latter part of it, under circumstances the most adverse and perilous. She was commanded by Captain Christie, formerly first officer of the "Great Liverpool," and who has had long experience in transatlantic, as well as coasting steam navigation.

We sailed from the Mersey on the morning of the 22nd of April, with little or no aid from the wind, and, though the engines were somewhat stiff, got abreast of Cape Clear in the afternoon of the 23rd. Had the weather remained favourable we should probably have reached Kilrush, on the Shannon, in 52 hours from the time of our departure; but, shortly after midnight, on the morning of the 24th, when we had left the light on the larger of the Skelligs Rocks a few miles astern, a gale sprung up from the southward and westward, and, acting on a previous swell from another direction, produced a remarkably heavy sea. In a few hours the wind rose to a perfect hurricane, and the waves, at once crossing and precipitous, rose to a fearful height, so that the hull and engines of the vessel were put to the severest trial. We stood to the north-west to keep clear of the land, which, though not very far distant, was invisible, even after daybreak, through the haze and spindrift. The spray dashed over the ship in continued sheets, and she encountered many tremendous seas that threatened to overwhelm her, taking her, indeed, sometimes on the weather bow and the lee quarter simultaneously; but such were her buoyancy and liveliness that she accomodated herself surprisingly to the emergency, and, without a perceptible strain, shipped very little water. She was put under the close-reefed mainsail and fore-stay-sail, but for some time during the height of the storm the boat cloths, placed against the weather main rigging, were substituted for the former. At one time, while, from the force of the wind, the height of the waves, and the frequent smothering of the paddle wheels, she was doing little more than lying to, a sea suddenly struck her and filled her main deck to a considerable height. To use the Captain's expressive phrase, she "sighed" under it for a moment, and then again rose like a duck. Another sea carried away great part of the larboard bulwarks forward, and damaged her paddle boxes, head rails, and figure head, but none of the iron-work was disturbed. During the gale, the arm of the mate was broken by a blow from a sheet-block, and three other men were disabled.

|| The "Erin-go-bragh" was built for the City of Dublin Steam Packet Co., and of about 330 tons burthen, with engines of 100 horse power, by Messrs. Mather, Dixon, and Co.

Till about ten in the morning of the 24th, the storm continued without cessation—the waves, which were steep and tumultuous, rising mountains high. About 8 o'clock, some of the men on board thought they saw a long ridge of dangerous rocks on the lee. If such were seen at all, (and there certainly was for a time an appearance to warrant the apprehension,) they were doubtless the Blaskett Rocks—to touch which would have been inevitable and instant destruction, but they were probably only the dark ridges of the waves indistinctly seen in the spindrift. Be this as it may, our gallant little vessel got clear of that ugly locality. The gale continued with occasional slight abatement during the whole day, and also the following night, during which it again increased to another hurricane, at (which is somewhat remarkable) the same hour. As the wind shifted some points from time to time, we occasionally put the vessel, which behaved admirably throughout, on the other tack, forging out seaward as we best could.

On the morning of the 25th, the weather became more moderate; and, though the sea was still running high, we put her before it under both sail and steam, and ventured, for the first time, to approach the land. She scudded admirably and safely. With all the disadvantages of her peculiar build, which gave her but little hold of the water, she had made a good offing; for we ran eastward about 50 miles, at about eleven knots an hour, before we again saw the Skelligs; the first "land" we made, if rocks may be so called. We immediately shaped a direct course to the estuary of the noble Shannon. We arrived off Kilrush in the evening, after a beautiful and interesting run from the Skelligs, and finally anchored at Tarbert, for the night, the engines never having once stopped during the whole passage.

The "Erin-go-bragh," after all the buffetting she had received, had neither sprung a single plate, nor started a single rivet. It was the opinion of those on board, as it is my own, that a wooden vessel could not have weathered it out, as her whole frame must have been so strained that she could not have been kept afloat. No vessel, whether of wood or of iron, with sails only—(and this I will venture to say in behalf of steam)—could have worked off so dangerous and proximate a lee shore, under the pressure of such a storm in wind and wave. I have several times crossed the Atlantic, and have encountered many gales; but do not remember to have beheld so trying a sea-way; for, in the open ocean the waves have room to roll away in their majesty and sublimity, without encountering the backset or recoil from the awful cliffs which here oppose them. The storm was severly felt on shore, many of the houses and cottages along the coast and for miles into the interior having been unroofed. A brigantine from Kildare, which had been in the tail of the gale, afterwards put into Kilrush dismasted, only seven feet of her main-mast being left.

You may believe, Sir, we were all grateful again to set foot on *terra firma*, and that, whatever our former impressions in behalf of wooden vessels, we had nothing to say against the "Erin." For myself, I am bound to add, that though my own slender nautical education was all imbibed in the old floating schools, with "wooden walls," and though my partiality was consequently strong in behalf of hard "heart of oak," my confidence in the still *harder* material for ship-building is firmly, and I trust not erroneously, established; and, taking all things into account, I am inclined to exclaim—"There is nothing like *iron!*"

After our arrival at Limerick, I made the passage to and from Kilrush in the iron steamer, "Garryowen", belonging to the same company. Though she had been running nearly seven years, she was as sound, and as apparently free from corrosion, as on the day on which she was launched, after having been all that time in a tide-way, without any repairs. Even the original red paint-work on her sides in the engine room was quite fresh and bright. She was taking in her third set of boilers during my stay at Limerick. The "Erin-go-bragh" is now running on the same station and is a great favourite.

I am, Sir, respectfully,

Your most obedient Servant,

John Grantham, Esq., Liverpool. JAMES SCOTT WALKER.

(No. 4.)

Liverpool, 19th March, 1842.

Dear Sir,—Understanding that you are about to publish a history of the application of iron to the building of boats and ships, and having been the first individual, I believe, that ever launched an iron boat on salt water, I take pleasure in communicating the particulars of the earliest attempts to introduce iron for ship-building purposes, and which, I trust, will not be altogether uninteresting.

In August, 1815, I launched a small iron boat, which I fitted up as a pleasure boat, and frequently sailed in it on the river Mersey. It was built by Mr. Joshua Horton, of Tipton, near Birmingham, but fitted up in Liverpool by the late Mr. Roger Hunter, and the late F. J. Humble. When not in use, this boat was put up in the Duke's Dock, where it was open to the gaze of any passer by; and, not being what a sailor would term *ship-shape,* owing to its being built inland, it was rather a curiosity. Its buoyant powers, however, and the remarkable ease with which it maintained its way, when once put in motion, attracted the notice of many. After having been a spectacle for the public for many weeks, this boat disappeared, and no trace of it could be discovered for a long time; but, on the next occasion of the Duke's Dock being let dry, it was found crushed up and deeply imbedded in earth at the bottom of the Dock. It is impossible that this boat could have been sunk without the use of some powerful mechanical aid, or of tools of some description; and the conclusion therefore is, that it was injured for some malicious purpose. The boat thus ruined was got up out of the Dock and sold for old iron, returning to me a fair proportion of its original cost.

The loss of this boat turned my attention to the practicability of making an iron boat which could not be sunk by any ordinary means, and, having matured my ideas on the business, I sought the means of putting them to a practical result; and for that purpose, in the year 1817, I made an agreement with the brother of Mr. Horton, the builder of my first boat, with the view of his settling in Liverpool, as an iron ship and boat builder. A suitable yard was taken, and a rent paid upon it for six months; but this project was frustrated by the death of Mr. Horton, by typhus fever, before he could remove to Liverpool. This disappointment checked my efforts for a time; but, determined not to let the subject rest entirely, and believing my invention a valuable one and worth while securing to myself, I lodged a Caveat in the Patent Office, in April, 1818, with a view of taking out a Patent, and thereupon proceeded to have a model constructed of a *life-boat*, to be built of *iron*, and which, from its peculiar form would possess to a great degree the property of righting itself, in case of being blown over by a squall of wind, and also of baling itself of any water it might ship from this or any other cause. This model, which, for convenience sake, was made of copper instead of iron, but which substitution of metal of course made no difference in displaying the principle of my invention, was exhibited in the Under-writers' Room at Liverpool, and its properties allowed to be tested in a tank of water in which it floated for that purpose.

The Caveat which I lodged in the Patent Office, I renewed several times; but, circumstances not favouring my again attempting the building of iron boats and ships for my own benefit, I at length sent my model boat to the same individual who had built my first boat, and from that model was constructed the second iron boat that ever floated on the salt sea, and which boat was a complete life-boat.

In September, 1822, I wrote a description of this life-boat, which was published in the *Kaleidoscope*, vol. iii., page 105, and a reprint of it was afterwards given by Mr. Egerton Smith, in his publication, entitled *Desultory Suggestions for Prevention from Shipwreck*; and as that description points out many of the superior properties of iron for the purpose of ship-building, which have since been practically proved to be true, although only yet, I believe, known to but very few, I beg leave to enclose you a copy.

The iron life-boat met with no better treatment than the first iron boat; for, having been lost for a long time, it was at length discovered imbedded in sand at dead low water of a remarkably low spring tide. The cause of its so lying being manifestly occasioned by some holes drilled into her chambers by some unknown hands. But this boat, having been found again, and not otherwise materially injured, was repaired and sold for use in the West Indies, where, no doubt, it is still doing good service to its owner.

I am, dear Sir,

Yours very truly,

John Grantham, Esq., Liverpool. THOS. JEVONS.

(No. 5.)

EXTRACTS FROM LETTERS TO MR. LAIRD.

Commander Cleveland, R. N., Captain of the Hon. East India Company's Iron War-Steamer, "Phlegethon."

" *Calcutta*, 5*th June*, 1841.

" We arrived at this place on the 22d of last month, as *sound* in *hull, boilers*, and *engines*, as when we left England. We have had some severe trials, and a large share of stormy, bad weather. I send you an abstract of our log, which will give you our track, &c.

"The 'Phlegethon,' in bad weather, has surpassed my most sanguine expectations, and, having gone over 17,157 *miles* without *straining a rivet*, will, I consider, quite carry out the principle that these sort of vessels can navigate in security between England and India."

Captain Hall, "Nemesis."

" *Hong Kong*, 21*st August*, 1841.

"I have just had her on shore again, to give her bottom the last touch before going north. *Nothing can look better—not a rivet started.*"

Captain Hall, "Nemesis."

"*Hong Kong*, 22*d August*, 1842.

" The 'Nemesis,' I am happy to say, is once more under weigh for active service, full of provisions, coal, and ammunition; the Admiral and fleet in company.

"I have had her on shore again, to make an improvement in her rudder, and nothing can be more *satisfactory than* the *examination I made of* her *bottom,—not a rivet started anywhere*, and our *machinery in as good order as when we left England.*

"In the late actions, the 'Nemesis' was engaged with the enemy 14 different times, and always in the hottest part of it. We are likely to be in the midst of it in a few days, and have every hope of giving you as good an account of our stewardship."

Captain Hall, "Nemesis."

" *Ningpo*, 1*st November*, 1841.

"We are all well, and anxious to make the most of our vessel, which looks nearly as well as when we left Portsmouth;—as for getting on shore, it makes very littte impression on her."

Captain Lynch, Commander of the Euphrates Expedition.

" *London, 4th April*, 1841.

" The 'Euphrates' has now been running under many disadvantages, in an extremely warm climate, often grounding with much violence, and receiving shocks that would have disabled an ordinary wooden vessel. She has never required any repair in the iron parts of her hull, or caulking.

" The wood-work has not stood so well, from the extreme heat of the climate; and it would be worthy of your consideration whether the decks and other portions, now of wood, could not advantageously be in future of iron. The rivetting has stood every shock without the least apparent change, and the vessel will run without requiring repair in her iron work for a longer period than I should like to predict."

Commander Charlewood, R. N.

" *July 1st*, 1841.

" I have no hesitation in saying that their strength" (iron vessels) " is greater in proportion to wooden vessels. The 'Euphrates' steamer met shocks in descending the river, by repeatedly striking upon banks and shoals, which a wooden vessel of the same size could not possibly stand; yet we left her at Bagdad in perfect condition, not a rivet started, nor a plate with more than a trifling dent on it."

(No. 6.)

" *Worsley, 24th Dec.*, 1840.

GENTLEMEN,—Some time has now elapsed since the iron steam boat " Alice," which you built for the Bridgewater Trust, was got to work; and it is due from me to inform you—which I do with much pleasure—that the expectations formed of her by the proprietary have been fully and satisfactorily realized, in her stability, light draft of water, and sailing qualities. She has, in every respect, answered the purpose intended in the trade in which she has been employed; and, in the event of another vessel being required for the same station, I would recommend her to be built of the same material and model.

The passage of the Mersey between Liverpool and Runcorn is often very difficult to make in stormy weather, of which we have had a more than usual succession in the present year; yet, notwithstanding this circumstance, and the tugging of from ten to fifteen heavily laden vessels at a time, as well as invariably (and unavoidably) taking the ground between tides, the "Alice" has withstood every strain, and has not sustained the least perceptible damage or derangement.

I am, Gentlemen,

Yours most respectfully,

EDMUND SMITH.

Messrs. Page and Grantham, Iron Ship-builders, Liverpool.

INDEX.

APPENDIX.

INDEX OF PLATES.

LIVERPOOL:

PRINTED BY LACE AND ADDISON, STANDARD OFFICE, NORTH CRESCENT.

KEELS.

Fig. 1.

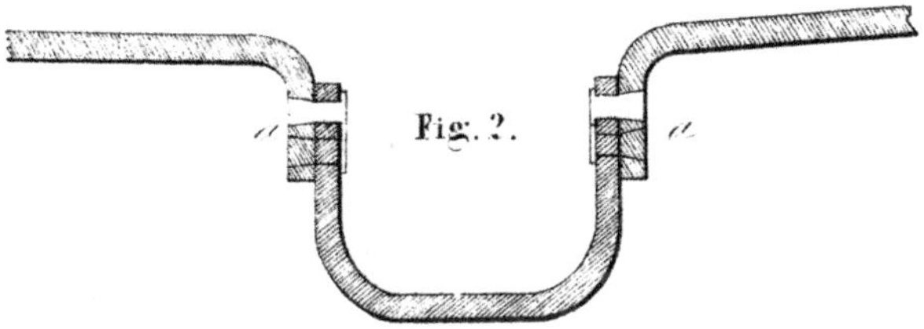

Fig. 2.

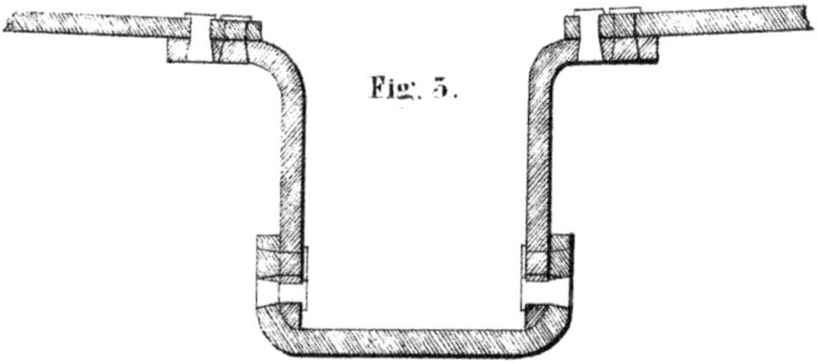

Fig. 3.

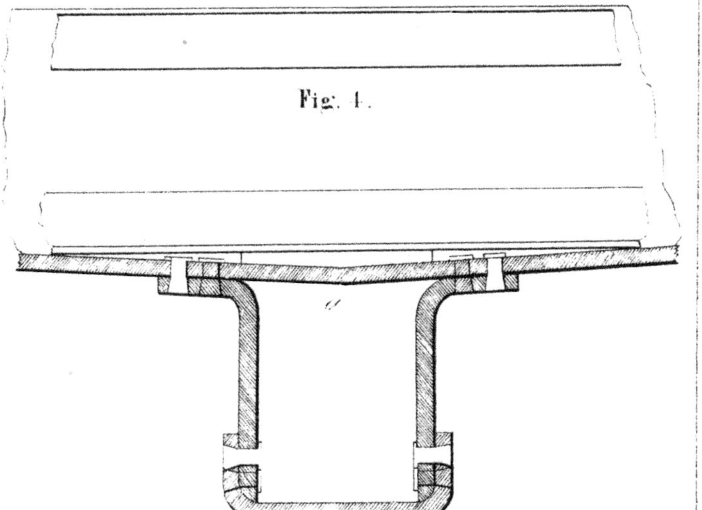

Fig. 4.

Plate I

STEMS.

Fig. 5.

Fig. 6.

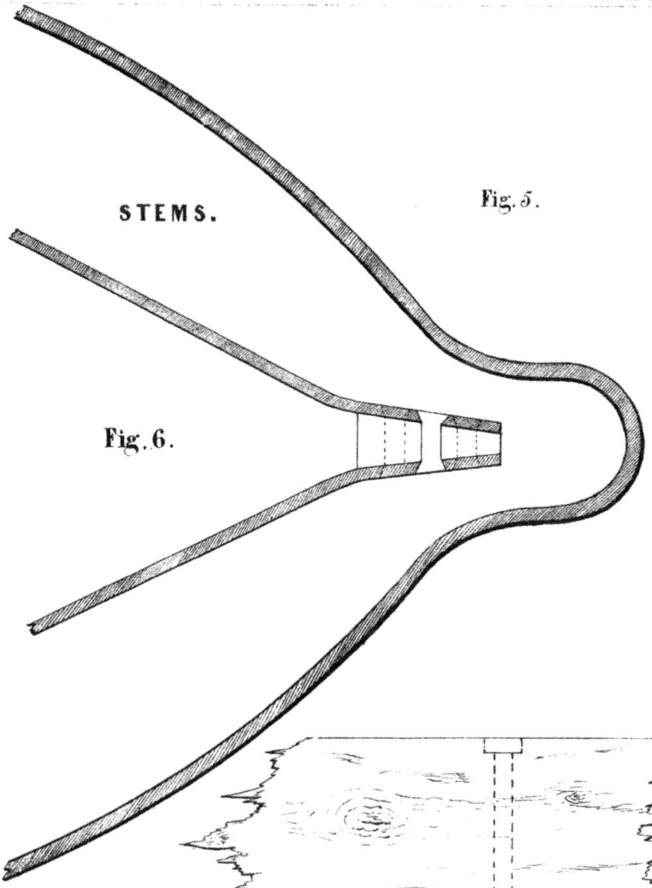

SECTION OF FLOORINGS.

Fig. 7.

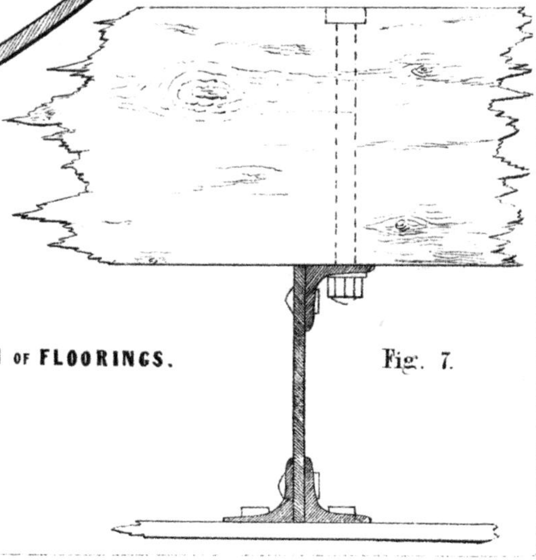

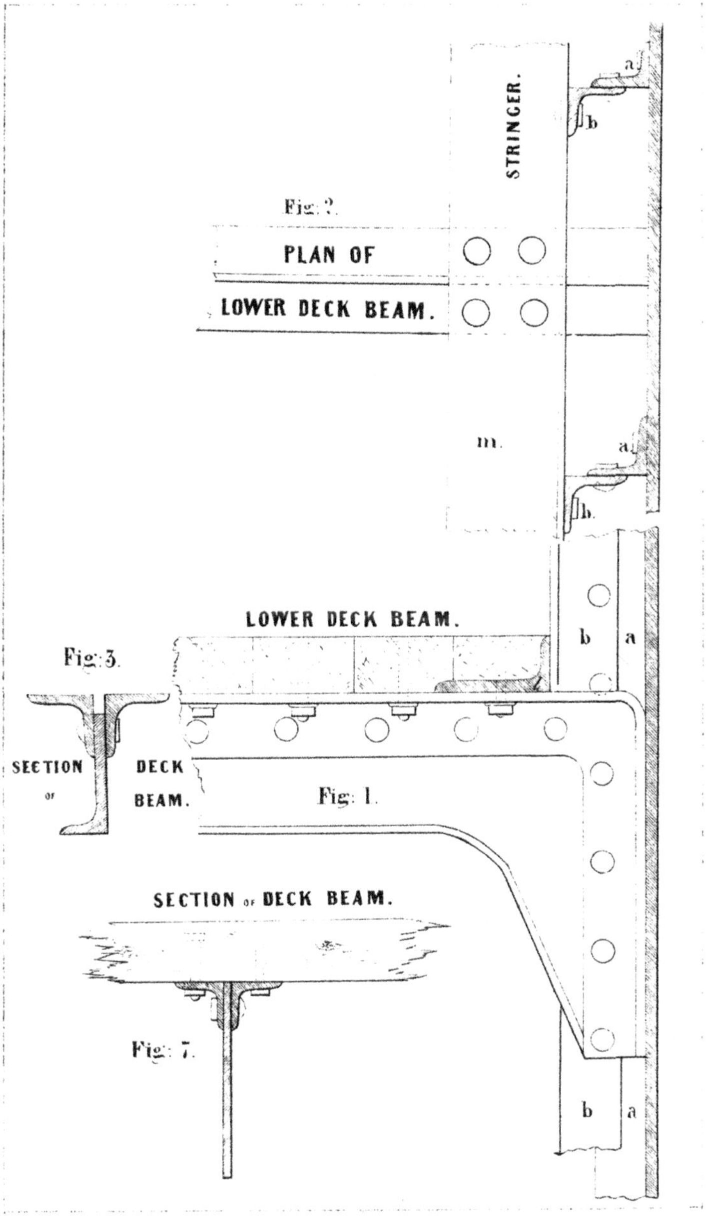
STRINGER.
a
b
Fig: 2.
PLAN OF
LOWER DECK BEAM.
m.
a
b
LOWER DECK BEAM.
b
a
Fig: 3.
SECTION
OF
DECK
BEAM.
Fig: 1.
SECTION OF DECK BEAM.
Fig: 7.
b
a

Plate 2.

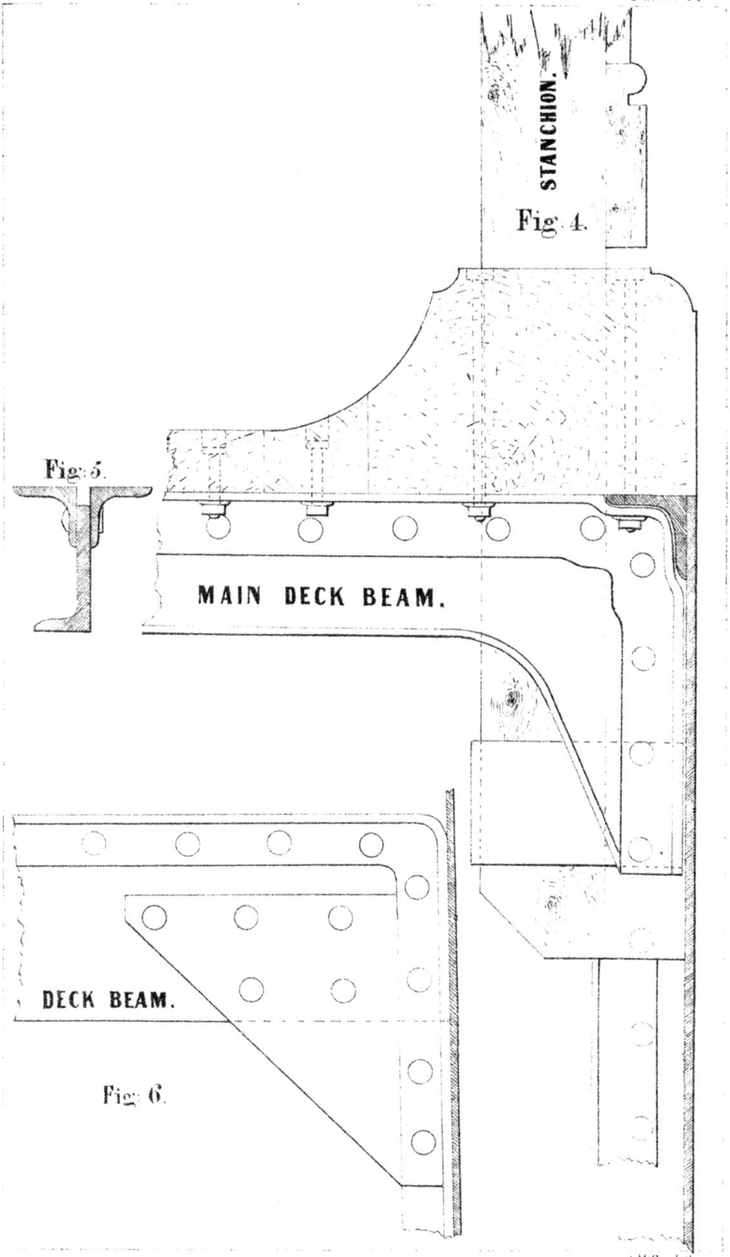

A Miller Lithog

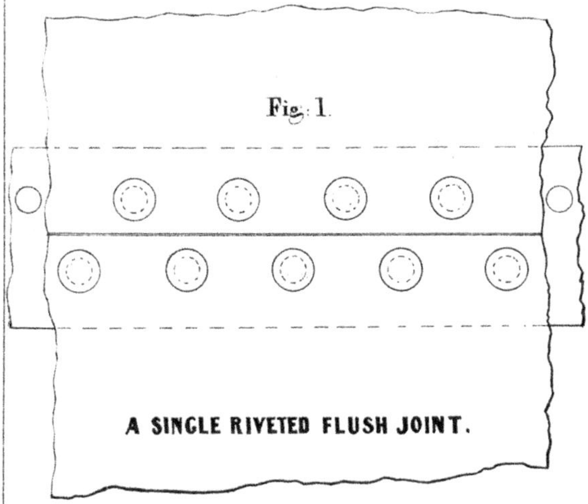
Fig: 1.
A SINGLE RIVETED FLUSH JOINT.

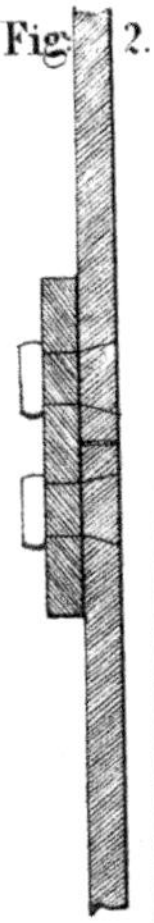
Fig: 2.

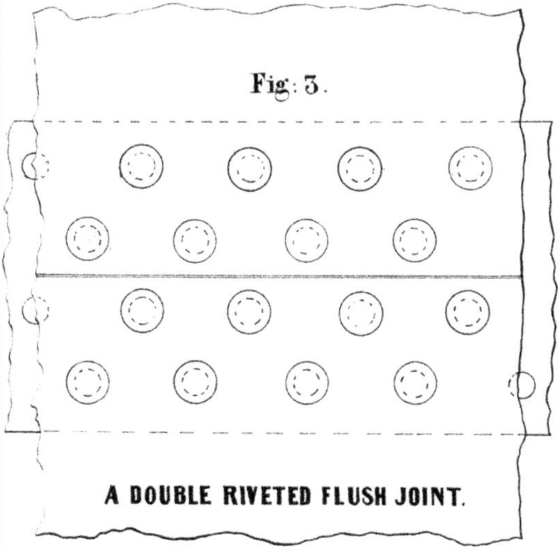
Fig: 3.
A DOUBLE RIVETED FLUSH JOINT.

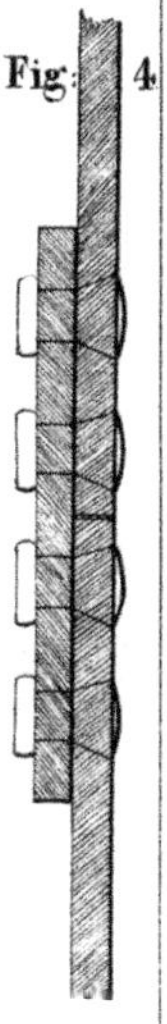
Fig: 4

Plate 3

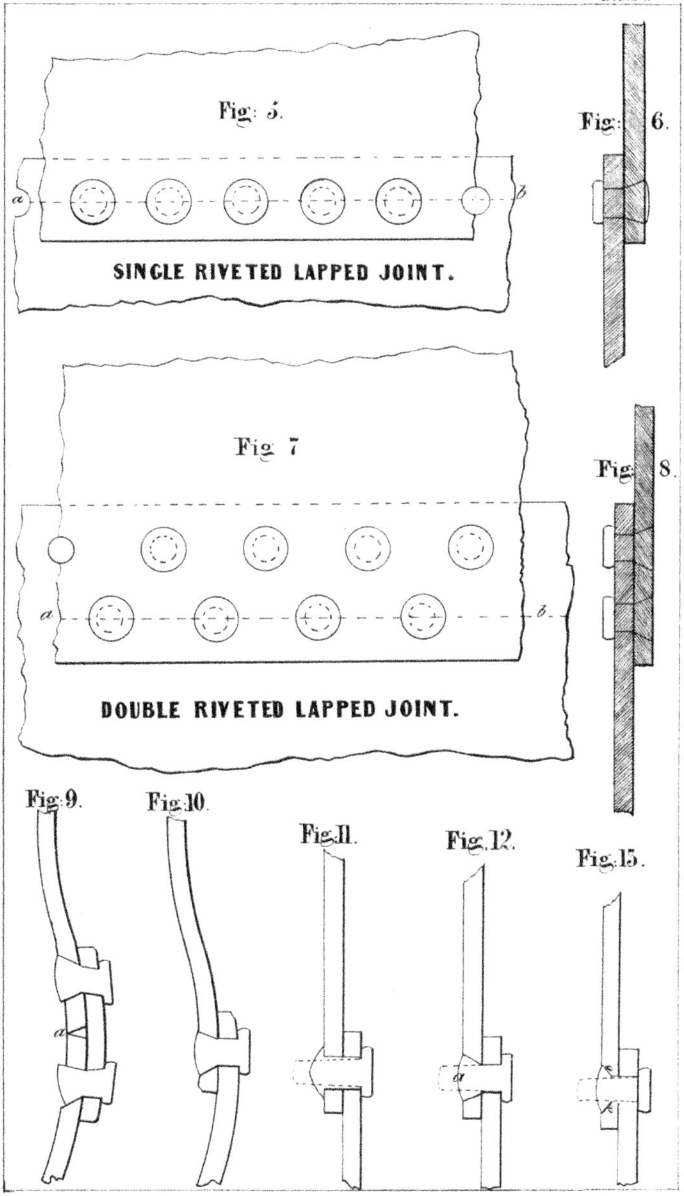
Fig: 5.
a
b
SINGLE RIVETED LAPPED JOINT.
Fig: 6.
Fig 7
a
b
DOUBLE RIVETED LAPPED JOINT.
Fig: 8.
Fig: 9.
a
Fig: 10.
Fig. 11.
Fig. 12.
a
Fig. 13.

Section of Plates,
Cut through the centre of the Rivets,
in the Planing Machine.
Drawn natural size.

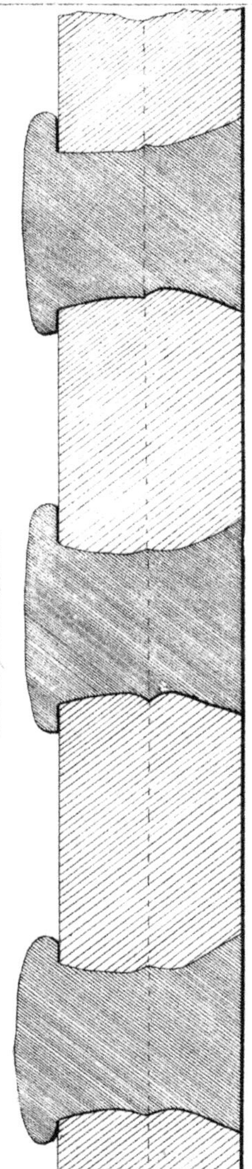

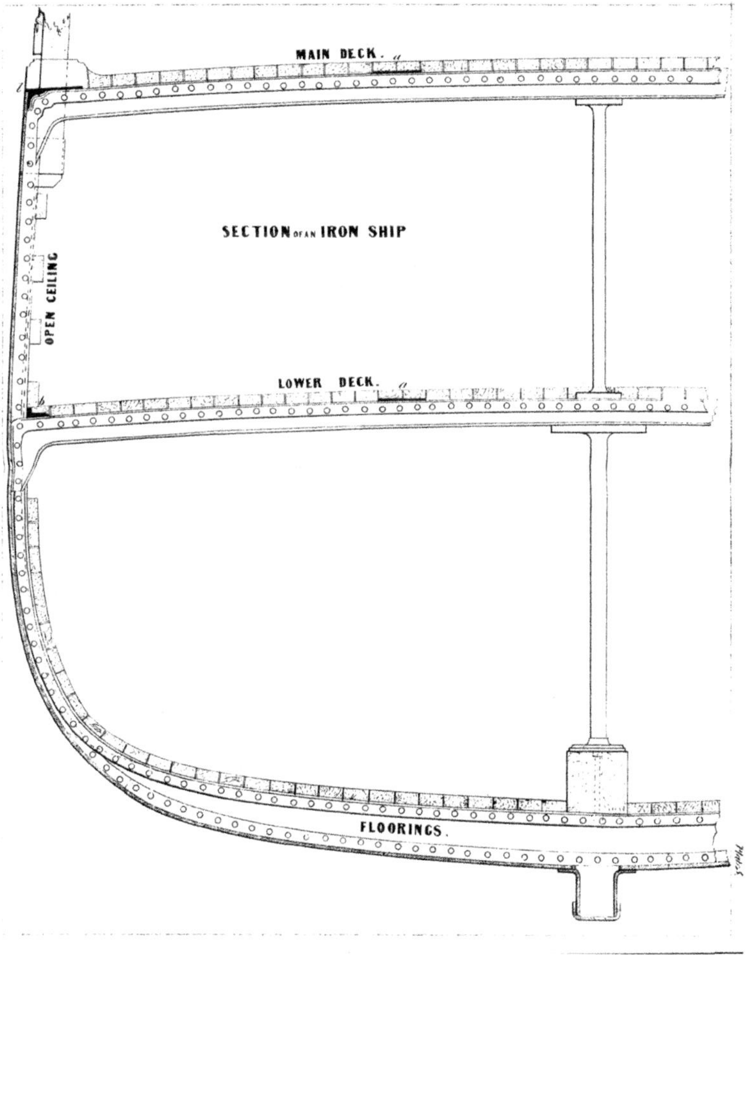
MAIN DECK.
SECTION OF AN IRON SHIP
OPEN CEILING
LOWER DECK.
FLOORINGS.

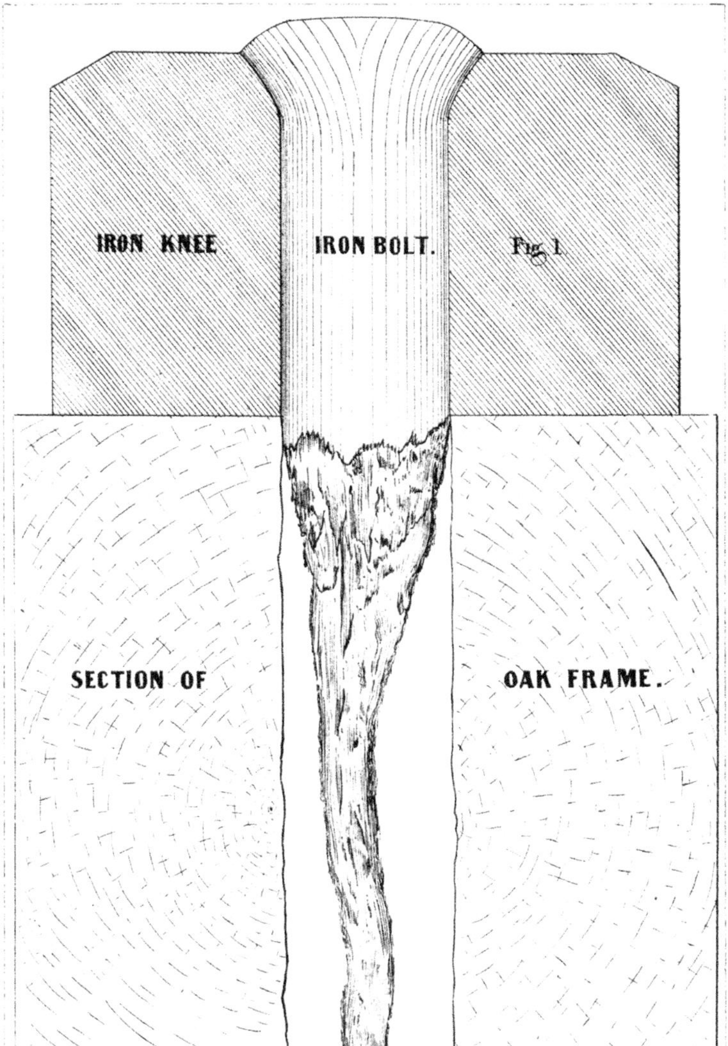
IRON KNEE
IRON BOLT.
Fig. 1
SECTION OF
OAK FRAME.

Plate 6.

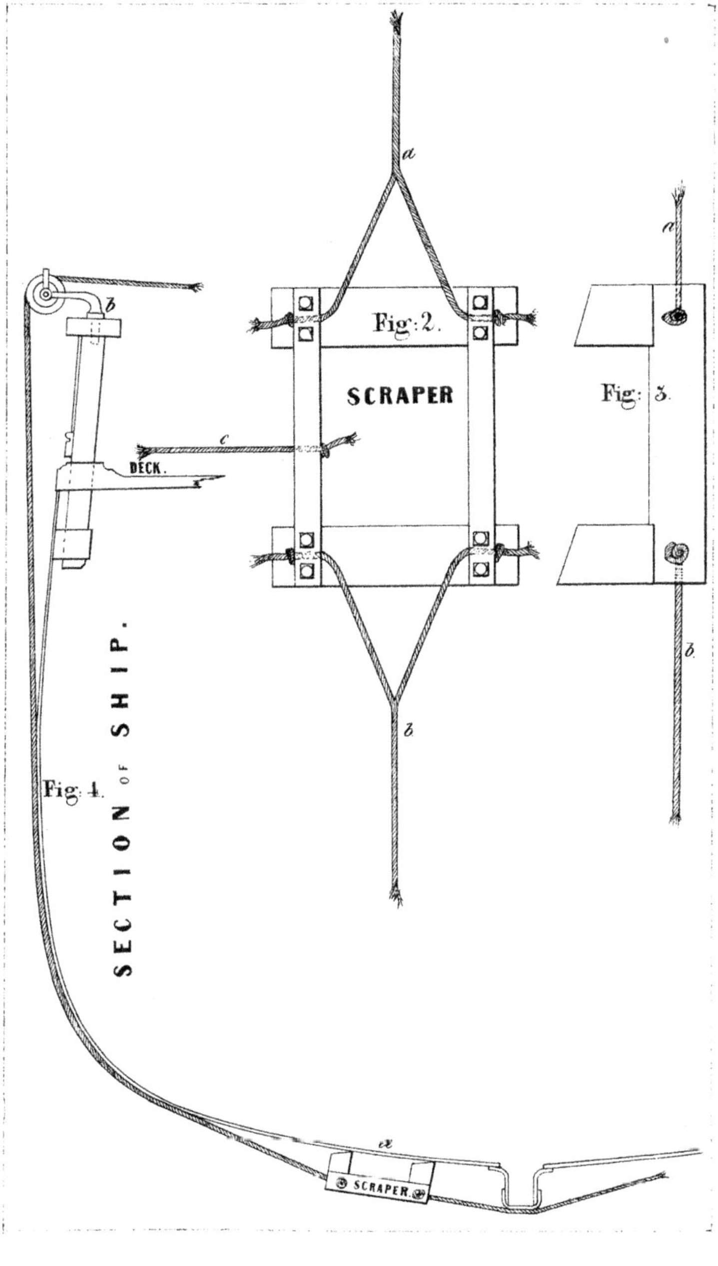

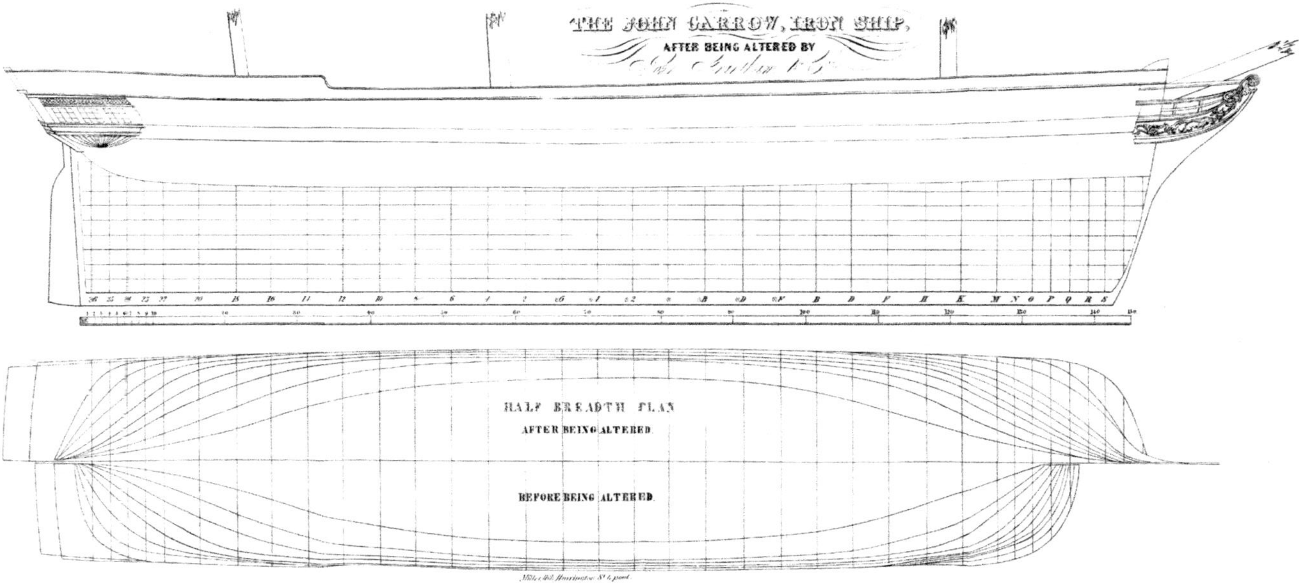
THE JOHN GARROW, IRON SHIP,
AFTER BEING ALTERED BY
HALF BREADTH PLAN
AFTER BEING ALTERED.
BEFORE BEING ALTERED.

Printed in Dunstable, United Kingdom

91221353R00070